KINGDOM OF FROST

BJØRN VASSNES

Translated by **Lucy Moffatt**

KINGDOM

— OF —

FROST

How the Cryosphere Shapes Life on Earth

GREYSTONE BOOKS
VANCOUVER/BERKELEY

Greystone Books Ltd.

greystonebooks.com

Cataloguing data available from Library and Archives Canada

ISBN 978-1-77164-454-9 (cloth)

ISBN 978-1-77164-455-6 (epub)

Editing by Dawn Loewen

Proofreading by Jennifer Stewart

Jacket and text design by Nayeli Jimenez

Jacket photograph by istockphoto.com

Printed and bound in Canada on ancient-forest-friendly paper by Friesens

Greystone Books gratefully acknowledges the Musqueam, Squamish, and Tsleil-Waututh peoples on whose land our office is located.

Greystone Books thanks the Canada Council for the Arts, the British Columbia Arts Council, the Province of British Columbia through the Book Publishing Tax Credit, and the Government of Canada for supporting our publishing activities.

This translation has been published with the financial support of NORLA.

CONTENTS

Earth's History

4,500 MILLION YEARS AGO (MYA): The Earth is formed

4,280 MYA: Water begins to condense in the atmosphere

3,600 MYA: The first supercontinent (Vaalbara) is formed

3,500 MYA: The first single-celled organisms, prokaryotes, appear; also, the first oxygen-producing bacteria

2,900 MYA: First glaciation (Pongola) occurs; possibly first snowball Earth event

2,400 MYA: The "oxygen catastrophe"; oxygen forms in earnest

2,400 TO 2,100 MYA: The Huronian ice age (with at least two snowball Earth events)

CA. 2,000 MYA: The first eukaryotes appear (first complex organisms with cell nuclei)

850 TO 635 MYA: Ice age (Sturtian-Varangian), with two more snowball Earth events

600 MYA: The first multicellular organisms appear

542 MYA: The "Cambrian explosion"; many new species appear

443 MYA: The supercontinent Gondwana becomes covered in ice; mass extinction of marine animals

420 MYA: First land plants appear, along with first fish with jaws (sharks), insects on land

252 MYA: Volcanic period; carbon dioxide in the atmosphere increases to 2,000 ppm; oxygen falls from 30 percent to 12 percent

251 MYA: Mass extinction; 90 percent of marine animals and 70 percent of land animals die out

199.6 MYA: The Jurassic (age of dinosaurs) begins

55.5 MYA: Episode of warming (PETM, Paleocene–Eocene Thermal Maximum); North Pole at 73 degrees Fahrenheit

50 MYA: India collides with Asia; the Himalayas are formed

35.6 MYA: Temperature falls 18 degrees Fahrenheit in the Eocene epoch

34 MYA: Ice forms on Antarctica

30 MYA: Australia and South America separate from Antarctica

3.9 MYA: *Australopithecus* appears

3.0 MYA: Ice cap in the Arctic forms

2.58 MYA: The Pleistocene, the most recent ice age epoch, begins

2.4 MYA: *Homo habilis* appears

CA. 200,000 YEARS AGO: *Homo sapiens* appears

125,000 YEARS AGO: Interglacial period

116,000 YEARS AGO: The last ice age begins

CA. 21,000 YEARS AGO: The last ice age peaks

11,600 YEARS AGO: The ice age (and Younger Dryas) ends; the Holocene begins

CA. 1350 TO 1850: Little Ice Age

CA. 1950: The Holocene ends, and the Anthropocene begins

Prologue

THE DANCE OF THE WHITE CAPS

WE HAVE ALL seen the famous photo taken from Apollo 17 in 1972. This picture of our planet, alone out there in endless space, taught us to think of Earth as our home, our *only* home, as something precarious and fragile that we needed to take care of. For the environmental movement, the photograph became almost iconic. The picture also gave us our perception of Earth as the "blue planet," because so much of the surface is covered in blue oceans.

But there is something this picture does not tell us, something we could have seen if the image of the Earth had been filmed from out there rather than just photographed. Not for just a few minutes, either, but continuously, throughout the entire year and—if it were possible—over millions of years. If that film were then played back at high speed, we would see a different image: we would see a planet in constant flux, the white caps at either pole expanding—over land and sea— and then shrinking again, in time with the seasons. When it was winter in the north, most of the landmasses would be covered in snow, which would vanish again when summer came. And likewise the sea, in both south and north: great, white, snow-covered expanses of ice spreading and shrinking, spreading and shrinking—back and forth in an annual dance.

If the film ran a little longer, we would also see other movements, following a more extended rhythm: in certain periods, less white would be visible, in others, a little more. And if the film were really long, we would see something astonishing. On occasions, the white cap would spread out across the entire planet, turning everything white. Earth becomes like a snowball. Not a single dark or blue patch in sight.

But the opposite also happens: for periods at a time, all the white vanishes—but always returns again. Sometimes slowly, other times quickly. Now and then, it seems to happen rhythmically, in steady cycles. But then the rhythm is interrupted. The white cap goes awry, or suddenly disappears. At the end of the film, as it approaches our own era, we see the rhythm becoming quicker, more intense. And as the film stops, we see that the white is shrinking once again, faster than ever before. It is so striking that we wonder what will happen when the film continues.

To understand what is happening down there, to grasp the rhythm of this dance, we must leave Apollo 17 and zoom in to the surface of this unique planet, so different from its duller siblings, Venus and Mars. They may be beautiful enough in the night sky, but they are monotonous and dead by comparison with our spectacular, ever-changing Earth. What causes this dance, and how does it manifest itself to earthlings? Could it be that they don't even notice it?

MELTING

OW DOES IT feel to stand inside a vodka bottle while the world melts around you? Not too bad, if the bottle is made of ice, is human-sized, and is the same one Kate Moss once stood in for a vodka ad. Pretty good, in fact, if you're at the ice hotel in Jukkasjärvi, northern Sweden, with its ice bar and barstools, its spectacular ice decorations and glasses (only cold drinks! says the bartender) made from blocks carved out of the frozen river and shaped by professional ice artists.

It was in the nineties that I experienced this. They let me come in and take a look around even though it was May and the hotel, actually closed for the season, was in the process of trickling back to the Torne river, only to be resurrected the following winter. Since then, around 50,000 tourists a year, many from Japan and China, have flocked to the same spot during the four winter months the ice hotel stands. Construction starts in November and it's ready for check-in by Christmas or New Year; by May, the melt is well underway. In the years since, imitators have emerged in both Finland and Norway, although they often use slightly simpler construction materials and are therefore known as "snow hotels." But even this isn't so easy nowadays: when the latest addition was due to open on Kvaløya, an island off Tromsø in northern Norway, in winter 2016/17, its launch had to be postponed to the following season because it wasn't cold enough. The thing is, winter is no longer reliable: we no longer know when it will come or go. The ice hotel in Jukkasjärvi has also faced this problem but now aims to fix it in a way that will enable the hotel to stand all year round: it will be kept cold using solar energy. After all, this is the land of the midnight sun, so in summertime, the sun can work around the clock. The tourists should certainly be happy with the combination of midnight sun *and* ice hotel.

The cryosphere,[1] the frozen part of our world, has become an exotic tourist destination, almost like a threatened animal species. As the cryosphere shrinks around the planet, tourists stream to the Arctic to experience these astonishing phenomena—ice and snow—while they still exist. Tourists now pay hundreds of dollars for things we used to be able to do for free as kids, like sleeping out in snow caves or having snowball fights. For most people, it's a matter of spending just *one* night at the snow hotel in a reindeer hide sleeping bag after traveling halfway around the globe to get there. Better-off travelers may prefer to experience ice and snow from the increasing number of cruise ships that offer trips to Svalbard, Greenland,

Patagonia, and the Antarctic. There, tourists can stroll on icebergs, greet the penguins, and chill their drinks with ice that is several thousand years old.

But for most of us who can't afford to spend thousands of dollars on a cruise to the icebergs of Disko Bay or the Antarctic Peninsula, time is running out if we want to feel the snow beneath our feet. Norway's most iconic celebration of its Constitution Day on May 17 involves a procession with flags and brass band from Finse up to the Hardangerjøkulen (Hardanger Glacier) in southern Norway. But according to glacier scientists, this will only be possible for a few more years. By around the middle of this century, Norway's highest glacier will be all but gone. And the same goes for many of the other, smaller glaciers, unless the warming comes to a sudden halt. Likewise, Norway's national sport of cross-country skiing is now under threat. Already, major ski races like the one at Holmenkollen, Oslo, can only be organized with the help of snow cannons, and cross-country skiers must make their way ever higher into the mountains if they want to feel real snow beneath their skis. Roller-skiing just isn't the same. What does this mean to a people whose identity has been defined by the frozen world? "As white as white is the snow" and "blue gave its color to the glaciers, that's Norway, in red, white, and blue!"—as it says in a popular song often referred to as Norway's second national anthem.[2]

Some would say it doesn't mean that much. Not all that many of us go skiing anymore. And plenty of skiing competitions now use artificial snow. Even the cross-country champion Thomas Alsgaard has said he expects the sport of cross-country skiing to die out soon owing to the lack of snow.[3] If the snow and glaciers did vanish here in Norway and other northern areas, we'd still survive. Even the tourist industry would certainly cope with it, because we still have the northern lights and the midnight sun—two attractions, which, fortunately, divide the year between them. So is there really any reason to make a fuss about this? Some people probably think it's a shame to have to abandon their skis in the basement, while others will be happy not to have to clear the snow anymore or pay for a snowplow to keep the road to their holiday cottage open. Others again will see it as a sign of the end of days or at any rate an indication of global warming. And perhaps they'll think it may make sea level rise a bit, causing problems for people on remote Pacific islands.

But for most northerners, these are trifling matters in a world that is changing in so many other ways. Terrorism, streams of refugees, and the automation of labor are more important concerns. What does it really matter if there's a bit less snow, a bit less ice? Even in Greenland, where people have used the ice as a hunting ground for millennia because that's where they could trap seals, many people think it's fine that the ice is melting, because it will open up opportunities for massive mineral wealth. And in Finnmark, the county in northern Norway where I grew up, few people will miss the road closures that can last well into May. Or the snow clearing. Just let it melt!

I used to think that way myself—I, who grew up in the Arctic, in a time of proper winters generally lasting eight or nine months of the year. For most of my childhood, I lived in Norway's coldest region, on the Finnmarksvidda mountain plateau. However, several harsh, snowy winters in Tromsø in the 1970s when we had to dig tunnels to our houses contributed to my decision to leave the region, and I moved to the much less snowy, but consequently much wetter, area of western Norway. And since it was still possible to go skiing even there as long as you went high enough up into the mountains, I didn't miss the white stuff.

It was only when I came to much more southerly latitudes, to places where snow never fell and temperatures never came close to freezing, that I discovered the cryosphere. It was on the populous, sunbaked, sweltering Indo-Gangetic Plain in northern India and Bangladesh that I came to understand how important the cryosphere is. Because what was it that kept people here alive during the driest and hottest parts of the year? It was the snow and ice in the mountains, far off in the Himalayas, which aren't even visible from the plain. When the rain no longer filled up the ever-dwindling rivers in the months before the monsoon, it was meltwater from the snow and glaciers on the Roof of the World that ensured the rivers never ran totally dry.

Almost nobody was talking about that back then in the nineties when I traveled around on the banks of the Ganges and its tributaries, making TV programs about the rivers and what they meant to people. Not that I was giving so much thought myself to what would happen if the glaciers vanished. But the glacial rivers from the Himalayas, Tibet, and nearby mountain ranges such as the Karakoram and the Pamirs sustain the lives of several hundred million people, well over a billion in fact if you include the great Chinese rivers that arise in Tibet. Later, I discovered that this is not a unique phenomenon: there are other places on the planet where snow and ice are also vital for keeping people, animals, and plants alive. This is particularly true of the countries around the Andes, where many of the largest cities are dependent on meltwater. Even fertile California is at the mercy of the cryosphere, as demonstrated recently when "the snows of yesteryear" ended up falling as rain and no longer served as a natural reservoir. So the Kingdom of Frost, the cryosphere, is vital for large swaths of the Earth's population, especially in places where most people have never even seen either snow or ice.

But it is also more than a reservoir. As I immersed myself in the cryosphere and its history, I discovered that its significance dates far back in time and is much greater than the history books tell us: the frozen world has been an absolutely determining factor in the way life has developed here on our planet. Over the ages, its fluctuations—the dance of the white caps—have shaped landscapes, life, evolution, and, to a great extent, human history. Even phenomena as diverse as our upright posture, the first fields of grain, the modern-day border between Norway and Sweden, steam engines, automobile traffic, and the skills of chess grand master Magnus Carlsen and javelin thrower Andreas Thorkildsen have all been influenced by the cryosphere and its fluctuations. Not directly, but through the decisive influence the cryosphere has upon the climate—as we are in the process of discovering today.

Anyone who—right now—is asking whether there's another planet like ours, with living creatures who go about wondering the same thing? Perhaps because the starry sky was so clear, thoughts like this occupied my mind so much in those days that I decided to study astronomy and physics when I grew up.

With the passage of time, though, my fascination ebbed away. Astronomy felt a bit otherworldly and I found other interests. When I returned to science, not as a scientist but as a communicator, what most preoccupied me were the mysteries of life. And not least the ultimate mystery: how life came about and how living organisms assumed ever more complex forms until, eventually, creatures emerged that were capable of pondering their own existence. Darwin became more important to me than Einstein, and the evolution of the human brain became more exciting to me than black holes. This preference also applied to my job as a science journalist, because the brain was still a newly discovered, unexplored continent.

So when NASA and other organizations began to report the discovery of Earth-like planets where there might be life, I was skeptical. Of course, the thought could stir your imagination: What if there really was somebody out there for us to talk to? But my reading about the development of life told me that we are the result of a series of almost impossible, or at any rate improbable, events. Life did not simply arise of its own accord, especially not complex life. This was something the evolutionary biologists John Maynard Smith and Eörs Szathmáry wrote about in *The Origins of Life*.[4] They described eight transitions or revolutions life had to undergo before creatures like us could come about, living beings it was possible to communicate with. And to get all the way to this point, it was necessary to undergo all the transitions: there were no shortcuts.

The first transition was the emergence of self-replicating molecules, which created copies of themselves. Even this is still a mystery to biochemists, but the assumption is that RNA (the slightly less complex relative of DNA) may have been the first stage. We do not know if this was how it happened, and self-replication requires a combination of two mechanisms: not just a method for the actual replication (copying), but also a means of acquiring the energy needed to carry it out. Life must therefore have emerged in the vicinity of an energy source. And remember, this was long before life's usual means of capturing energy, photosynthesis (which converts solar energy to biological energy), was "invented." Some scientists, like biochemist Nick Lane, have therefore argued that the first living organisms must have arisen in or close to submarine hot springs or volcanoes.[5]

I won't go through all eight stages proposed by Maynard Smith and Szathmáry, or Lane's version of the development of life. Suffice it to say that it is theoretically possible to provide an explanation of how life on Earth evolved from simple, single-celled organisms, somewhere between 3.5 and 4 billion years ago, to more complex beings. That is not to say there is perfect clarity about all of the steps.

The story I will try to tell here—in a very short, simplified version—is how this development and life's different revolutions have been intertwined with the history of the cryosphere.

The connection appears to have been there from the outset. It all started several billion years ago with a chunk of ice that came sailing through space and collided into a blazing hot

Earth. This chunk of ice was a comet, and it was followed by a whole swarm of other comets and various celestial objects during the highly unstable early phase of our solar system's history. These celestial objects brought many things with them—of which more later—but one crucial contribution was the substance that actually forms the cryosphere: water.

Because water is what it's all about. Water in its many frozen forms: transparent ice, clear as glass, on the lakes; slop and slush on the roads; fern frost on frozen winter windowpanes; snowflakes drifting slowly through the air; compressed crystals beneath thousands of feet of glacier ice; black ice that suddenly springs up on the road sending cars into ditches; rime on withered straw in October; old spring snow that makes it impossible for animals and humans to get about; icebergs that strike ships in the night, sending hundreds of passengers out into the waves. And snow and glaciers that store water through the spring and melt in time to allow thirsty humans and beasts to drink. This is what makes the Earth unique: that we, here, can find water in all these strange frozen variants.

Water is an unusual substance, and all the more remarkable when it freezes. It isn't a question of magic but of water's physical properties, which result from the water molecule's distinctive form. This form creates especially strong bonds between water molecules, giving the substance unique properties, especially in frozen form. Water molecules are formed of hydrogen and oxygen atoms, which are bonded in such a way that the two hydrogen atoms attach to one side of the oxygen atom. This makes the water molecule "lopsided," giving it strong polarity, with a positive charge on the side where the hydrogen atoms are and a negative charge on the oxygen atom's side. This polarity creates powerful bonds between the water molecules, binding them together tightly in a "bent" form in a liquid or gaseous state, and as symmetrical, hexagonal crystal structures in a solid state.

These crystals, which can vary dramatically in shape but are mostly hexagonal under normal conditions, are bonded in a way that gives water several remarkable properties—among others, that of being lighter in solid than in liquid form, which is why ice floats on top of water. This property is shared by only a few other substances, including diamonds, which are actually a form of carbon. Under the right temperature conditions—on another planet or moon—we might see "icebergs" of diamonds looming up from a sea of liquid diamond.

However, we will never see this on Earth. Where we live, water is the only substance that can occur in all three states, solid, liquid, and gas, under conditions we can live in. Indeed, the three states can actually occur at the same temperature—32 degrees Fahrenheit, or 0 degrees Celsius (ice and snow can evaporate directly, without taking the "detour" via liquid water). This is because the strong bonds between the water molecules make it difficult to separate them, which gives water unusual boiling and freezing points. In thermodynamic terms, water is described as being extremely resistant to phase change. It takes a great deal of energy to melt ice into water, and also to make water evaporate. Water's special structure in frozen form, especially when it occurs as snow, gives it other unusual properties: it becomes white and light when it freezes, and snow is one of the substances that best retains heat. This is why you can sleep in a snow cave without freezing to death.

But in the earliest days of Earth's history, there wasn't much snow or ice to be seen. After Earth came into existence, during the turbulent beginnings of our solar system some

4.5 billion years ago, our planet was a ball of fire with a temperature of over 14,000 degrees Fahrenheit—hotter than the surface of the sun is today. Bombarded by a constant rain of comets, meteors, and other celestial objects, it was truly hell on Earth. Gradually things calmed down. After half a billion years, the gravitational fields of the sun or the planets had drawn in the solar system's stragglers, which had either landed or settled into a stable orbit, like the asteroids we might come across between Mars and Jupiter. Earth had begun to cool and now at last it could enjoy a gift brought here by all this bombardment. As I've noted, the comets and rocks had brought with them water, that singular substance with its unique properties on which we are so reliant. And not just water: scientists have now discovered that comets may have brought with them everything that is needed for life to emerge, perhaps even life itself—all bundled up in a packaging of ice.

What does it take to create life? First of all, there must be complex organic molecules, such as amino acids (the building blocks for proteins), nucleobases (the building blocks for genetic material), and carbohydrates. One of the prerequisites for the emergence of life is the presence of such molecules. But scientists do not believe these complex molecules existed on Earth at the time when living organisms are supposed to have appeared here. So how can life have emerged? One possible explanation, recently backed up by observations and experiments, is that these types of molecules actually came tumbling down from the heavens, from outer space. And they were apparently brought here by large chunks of ice, comets. If this is true, all of us have our origins in ice.

The idea that life came from outer space is not new in itself; indeed, it is so widespread that it has a name: panspermia.

Renowned scientists such as Francis Crick and Enrico Fermi have written about this, and it is a familiar theme in books and films.[6] Panspermia comes in different versions. One is that someone intentionally sent these "seeds" to Earth. Another is that living organisms survived their journey through space and landed here by chance. It has, in fact, been proved that certain tiny animals called tardigrades or water bears can survive such conditions. Scientists have tested this theory by sending them into space, where they go into a kind of hibernation but can be woken up again afterward.[7]

A more sober version is the one supported by recent discoveries: that what came to Earth with the comets was not living organisms but the building blocks of life. And precisely these types of building blocks have now been found on a comet, 67P/Churyumov–Gerasimenko, which has been extensively studied using instruments on the Rosetta space probe. The substances found to date are the amino acid glycine and the mineral phosphorus, which is also a necessary ingredient in living organisms. In addition, comets and other celestial objects must have brought water—also absolutely essential for life—to Earth, in the form of ice.[8]

According to Kathrin Altwegg of the University of Bern, a lead scientist on the Rosetta project, this shows that comets may contain everything that is needed to create life apart from energy (it is too cold on a comet). It is hardly likely that the glycine originated on the comet itself; it probably came from dust clouds that existed before the solar system was formed. The dust particles were, in fact, a good place for organic molecules to be formed, as demonstrated in laboratories. At that time, however, Earth was too hot for such fragile amino acids to be able to come into being here. What Earth could contribute, though, and what the comets lacked, was the energy needed for life to emerge from these organic molecules. They needed heat to begin to react with each other. This was why the encounter between frozen organic molecules and the heat of the Earth may have been what kick-started life.

These sorts of "start-up packs," or perhaps even frozen single-celled organisms, may have arrived on Earth early on, via comets or other celestial objects. We know Earth was heavily bombarded by such objects in its infancy, and water must also have been involved in this bombardment: we know there are still celestial objects out in space, like the asteroid/dwarf planet Ceres, which have large quantities of frozen water. Just a few collisions with such celestial objects would be enough to provide Earth with all the water we have today.

But it took a long time for these "start-up packs" to be opened. About a billion years had to pass before the conditions were ripe for life to develop here. The surface of the Earth had to cool, and the steam had to condense and fall as rain, allowing liquid water to form on the Earth's surface, and eventually oceans. Because the ocean is where life began.

Not only did life's building blocks come to Earth with ice, but the ice that had melted in its collision with the blazing world had to return, as the cryosphere, in order for life to begin developing here. Life and the cryosphere appear to have tracked and influenced each other through billions of years, although their dance was a very slow one in the earliest days. And it took a long time for the Kingdom of Frost to send its first snowflake down to Earth.

— 3 —

THE FIRST SNOW

WHEN DID THE first snow fall? No, not the kind that falls one day in November only to melt as swiftly as it came. I mean the *very* first snow here on Earth. The first snowflake to come drifting down upon an unprepared Earth, which had no idea this singular phenomenon would recur each winter. This first flake would be joined by many others, so many in the end that some remained on the ground throughout the summer, marking the start of the cryosphere, the frozen part of Earth.

Snowfall is light and quick to vanish, so how can we say when the first snow fell? It is difficult to establish a precise date. Nobody was there to witness it. In the unlikely event that it left any trace fossils, we wouldn't know where to look for them anyway because the continents have shifted so much over the ages. But what we *can* say something about is when snow first lay so long on the ground that it became ice, a glacier that created identifiable striations in the bedrock. We can see these striations and we have ingenious methods for dating them. It's all to do with the fact that the rock contains certain radioactive variants of minerals (isotopes): based on the amount of radiation they emit, we can calculate their age—take a reading of when they were formed. This is because we know the half-life of the different isotopes (how long it takes for their radiation to diminish), just as we can tell a glass of beer has been standing around for a long time when there are almost no bubbles left in it. If the minerals are magnetic, the scientists can even find out how they have moved by checking the direction of their magnetism. Earth has a magnetic field that creates patterns in certain minerals.

No snow fell in the earliest part of Earth's history, that much is certain. As noted earlier, Earth was hotter then than the sun's surface is today. But as the solar system calmed down, Earth had some respite from the bombardment, and the steam that had gathered in the atmosphere began to cool, falling as rain. Rain has a chilling effect, as we can tell when there's an afternoon shower on a hot summer's day. And gradually, Earth became cooler. Increasing amounts of water made the transition from steam to liquid form, creating pools of water, then lakes, and eventually an entire ocean. At the same time, the amount of steam in the atmosphere diminished, thereby reducing the greenhouse effect, which we now know keeps Earth warm.

The Greenhouse Effect

The much-discussed greenhouse effect is so called because it is reminiscent of the conditions in a greenhouse, where the heat of the sun *enters* through the transparent walls of glass or plastic, but less of it gets *out*. This is because the radiation that is reflected has a different wavelength than the radiation that enters. The Earth's atmosphere operates similarly, and the changing concentration of the different gases in the atmosphere determines how strong the greenhouse effect is. The greenhouse effect makes the temperature on Earth higher; without it, the planet would be at least 63 degrees Fahrenheit cooler. We can see the consequences of a strong greenhouse effect on Venus, where 96.5 percent of the atmosphere consists of carbon dioxide (CO_2) and the surface temperature is 872 degrees Fahrenheit.

The most abundant greenhouse gas on Earth is, in fact, water vapor, but the gases that are considered to be the strongest drivers of the effect are CO_2 and methane (CH_4). Studies of tiny air pockets left in ice and rock show that the concentration of these gases has fluctuated considerably over the course of Earth's history. Today, we believe the main source of the greenhouse effect is carbon burning, but this is only part of the truth—and was, at any rate, not the case for the first billions of years. Back then, these gases were emitted from the Earth's core, through volcanoes and other similar "vents." And since volcanic activity may have been high in Earth's earliest phases, the concentration of greenhouse gases was also high. This is probably the reason there was no ice age in the first 1.6 billion years, even though the sun's radiation was weaker then and Earth therefore "ought" to have been cooler. So the greenhouse effect is nothing new; it has, at times, been far more severe than it is today. Whereas CO_2 concentration has just passed 400 ppm (parts per million) and was 280 ppm in pre-industrial times, in previous periods it has been as high as 7,000 ppm. The reason we worry about the greenhouse effect today is its historically rapid increase, and the fact that it already appears to be affecting the climate in ways detrimental to life.

OUR NEIGHBORING PLANETS, Venus and Mars, offer us a good illustration of the significance of the greenhouse effect. On Venus, the greenhouse effect has run riot, causing the planet's surface to become insanely overheated and making it an absolutely impossible place for living organisms to survive. Mars has gone to the opposite extreme: it has almost no atmosphere—possibly because the planet is too light to retain one—and therefore also has no greenhouse effect. Here, the *mean* temperature is –76 degrees Fahrenheit, hardly propitious for life.

But back to Earth, which has been unusually fortunate in avoiding these extremes, partly because we are just far enough away from the sun and partly because we have acquired an atmosphere that provides just enough greenhouse effect. It would be more than a billion years before it became cool enough for the water molecules not just to condense and fall as rain, but to build one of nature's wonders: snow crystals, unique structures consisting of around a hundred quintillion water molecules. One reason why this took so long, even though radiation from the sun was so much weaker than today, is that water vapor is an efficient greenhouse gas. So as long as there was a lot of steam in the atmosphere, Earth was very humid and hot. But as the steam gradually cooled, condensed, and fell as rain, the temperature dropped enough to allow the water to freeze at last. Finally, after a good billion years, the first snow could fall.

Snow is born high up in the atmosphere. It mostly forms in one of two situations: either when humid air comes in from the sea and is pushed upward as it meets a mountain range, such as the Cascades of western North America; or when warm and cold air masses meet. In both cases, the warm, humid air is pushed high up into the atmosphere where it cools; if dust particles that can serve as a nucleus are present, the vapor begins to form snow crystals. This does not automatically happen at the freezing point: snow may also form at higher temperatures. And the opposite is also true, as many drivers know from bitter experience: rainfall may be supercooled, below zero, which causes it to freeze as soon as it hits the ground. In the dialect of Hardanger, western Norway, it's known as a *juklasprett*, which translates roughly as "glacial bloom": you can literally see how the ice, the *jøkul* or glacier, springs up from the ground, sending cars off the road and making people fall and break their hips.

As we've seen in countless illustrations, snowflakes can occur in an infinite number of forms, depending on conditions such as temperature, humidity, and wind—as well as sheer, simple chance. The crystals may look like stars, polygons, discs, pillars, or plates. They may be more or less loose or compact. A common shape in dry conditions is hexagonal. It is reflections from such crystals that can create "sun dogs" or "moon dogs," bright spots on either side of the sun or moon.

Like other crystals, snow crystals have a tendency to build up, and when they are heavy enough, they start to fall. They often melt before landing or as soon as they come into contact with the ground. But if it's cold enough, they may settle. And if more snow falls, it can begin to accumulate. The length of time it remains depends on the temperature and how much snow manages to fall. Snow may come and go many times before settling for good. Sometimes not even the summer sun can thaw it all, and "the snows of yesteryear" remain until the next winter, in a much more densely packed form than they had to begin with. This is how glaciers are born.

Snow crystals have many relatives: sleet, or raindrops that freeze partway to the ground; hailstones, which have never been snowflakes but started out as shapeless particles of ice formed around a frozen core in cumulonimbus clouds; graupel, snow pellets that form when supercooled water droplets freeze on falling snowflakes; hoarfrost, the frozen version of dew; and rime or glaze, the coating of ice that results when supercooled fog or rain comes into contact with objects (black ice when it covers roads). Water can take on countless forms when it freezes. These forms are not just beautiful but also useful. I've already mentioned the insulating properties of snow. But both snow and ice have several important functions. If they remain, they keep water in place, ensuring that it doesn't run off at once but is stored either temporarily or permanently as snow, frost, or glacier ice. Sometimes, it may be stored for just a few days or weeks or through the winter. Other times, it can remain there for very long periods, lasting thousands of years, until a warmer climate releases the water once again. These fluctuations in the frozen world, over different timescales, create a dynamic that has shaped not just our landscapes but also life itself.

But when it comes to long-term history and the major fluctuations in the cryosphere, one property of snow in particular is vital: its whiteness. This gives it a thoroughly unusual capacity to reflect sunlight, up to 90 percent of sunlight in the case of new snow. If we take into account the fact that snow can cover up to half the land surface of the northern hemisphere in winter, as well as large expanses of sea ice and glaciers, it goes without saying that the climate effects can be considerable. Indeed, this albedo effect (see "A Closer Look: Albedo") can trigger self-reinforcing climate processes in both directions: when the albedo diminishes because the sea ice and snow cover are vanishing, the temperature rises because land and sea absorb more of the sun's heat, which leads to even more melting, and hence more heat absorbed, and so on. The opposite also applies: when more snow comes, the albedo increases, more of the sun's heat is reflected, and it grows even colder, and so on, in a self-reinforcing feedback mechanism that has triggered an ice age on several occasions.

SNOW AND ICE also affect the climate in other ways, albeit at a more local level. When water freezes in the autumn, it releases large amounts of energy, which has a warming effect on the surrounding area: it feels warmer than it actually "ought" to be. In spring, when snow and ice melt, the opposite occurs. Melting takes a lot of energy, which makes the air grow colder than it would otherwise be. So in both autumn and spring, snow works as a kind of buffer, making the temperature changes happen a bit more slowly than they otherwise would. And snow has even more unusual qualities owing to the special structure of the snow crystals. One of them is that snow, though cold in itself, is one of the best insulators of heat in existence. This is what makes it possible for reindeer to find unfrozen lichen beneath the snow—and what makes snow-free winters a nightmare for reindeer herders, my neighbors during childhood.

Albedo—the Effect of Whiteness

One reason why the cryosphere is so important for the climate is its whiteness, *albedo* in Latin.[9] The Latin word is used to describe how much of the sun's radiation a surface reflects. The reflection depends on the wavelength of the radiation, the angle at which it strikes the surface, and the nature of the surface. How much of the sun's energy the surface of the Earth reflects has major implications for the temperature.

When snow settles on the ground, the albedo increases. New, dry snow reflects between 80 and 90 percent of the radiation. We say that new snow has an albedo of 0.8 to 0.9, where 1 indicates full (100 percent) reflection. When snow has been on the ground for some time and has become compacted and dirty, the effect diminishes but will still be considerable.

Sea ice has an albedo of 0.5 to 0.7, while open sea has an extremely low albedo of around 0.06. This means that when ice forms on the sea, the albedo increases dramatically, even more so if it is then covered in snow. While open sea absorbs almost all the energy from the sun and is warmed up, snow-covered sea ice reflects almost all the energy. With less ice and more open sea, the ocean will absorb more solar energy, which will cause even more ice to vanish, leading to more heat absorption, and so on.

Vegetation also influences albedo. Coniferous forests have almost no albedo, between 0.08 and 0.15. Deciduous trees have between 0.15 and 0.18, while green grass has an albedo of around 0.25. Generally speaking, the more forest and shrubs there are, the weaker the albedo effect; the more grass, the higher the albedo.

How important is the albedo effect? Today, Earth's mean temperature is 59 degrees Fahrenheit. Calculations have shown that if Earth were entirely covered in sea, which has a pretty low albedo (0.06), the mean temperature would be just below 80 degrees, which would make large swaths of the planet uninhabitable. On the other hand, if Earth were totally white, with an albedo of close to 1, the mean temperature would fall to around −40 degrees.

— 4 —

IN THE REALM OF THE SNOW QUEEN

SNOW IS QUIET. Not just when it falls, but also when it has settled and covers the landscape as far as the eye can see. Like on Finnmarksvidda on a winter's day, far away from all the houses, roads, and traffic—the way it could be on the plateau when I grew up there in the sixties. There was a silence that was more than the mere absence of sound. Because there *was* sound: the sound of silence, but most of all the sound of endless space. And of endless time, as if the snow had always lain there peacefully. Which, of course, it hadn't. Because snow also has another face.

At dawn, as we reached Bæskades, a storm came blowing up.
A flaying rush of driving snow whined on the mountaintop.
Heavily leaned we into the storm, forced to rest a while,
Our reindeer too were weary after many a long mile.[10]

It isn't so long since people traveling in northern Norway in midwinter had to go by reindeer over the Bæskades plateau, just like Nordahl Grieg. As depicted in his poem, idyllically entitled "Morning on Finnmarksvidda," it could be a grueling experience if the weather gods didn't smile upon you. A journey that takes two hours by car today in summer conditions could well take one or two days, so it was a good thing there were plenty of mountain lodges en route.

When I traveled over Bæskades in winter as a lad, it wasn't by reindeer but by a peculiar weasel-like vehicle known as a *snowmobil* (not to be confused with modern snowmobiles). It was a sort of tracked vehicle in which around ten passengers sat huddled together in a circle, barely catching a glimpse of the white landscape speeding past on the other side of a few tiny round windows, like the portholes on a boat. The *snowmobil* traveled faster than the reindeer and didn't tire as easily; it was usually the passengers who had to stop for a break and a drop of coffee. I don't remember how long the trip took, but it was the better part of a day. And why did I spend two days—there and back—on a trip like that? To stand shivering for hours on end watching somebody go around and around on a skating rink, that's why.

I was ten years old and lived in Kautokeino, Norway's most isolated municipal center, especially in winter. It was also the coldest—in competition with Karasjok—with winter temperatures sometimes falling toward –58 degrees Fahrenheit. That was fine by me: when it fell below minus 40, we were given the day off school. We moved to Kautokeino at the end of the 1950s, before the coastal town of Alta could be reached by a road that was open all year round, and when our community could still be isolated for weeks in spring. Then it was impossible to drive either car or *snowmobil* owing to the spring thaw. The snow grew wet and impassable, rivers and lakes could no longer be crossed, and even the reindeer had to throw in the towel. Today, a situation like that would merit helicopter airdrops and a news feature on TV, and the parliamentary public safety committee would be hauled in for a hearing to find out who was to blame. In those days, it was just the way the world was. The seasons had their rhythm: the snow came in autumn, the water froze, and later, in spring, everything started to thaw again and you just had to stay where you were, hoping you had enough of the bare necessities. It was what we were used to.

Young as I was, I knew no better and thought this was quite normal. As a child, I also got out of doing the most unpleasant chores, like going outside to fetch water from the ice-covered tarns or brooks when the water pipes froze. This was the kind of thing my father often had to do, and once he lost his footing and fell into the water. He ended up under the ice and spent a good while lying there flailing about before hauling himself out, soaking wet in the bitter cold. That he got home without freezing to death and didn't fall ill tells you something about how hardy his generation was up there in the north. The same must be said of the young woman who set off to give birth at the clinic in Kautokeino one cold winter's day—alone, on foot, in the snow. She didn't make it in time and had to deliver the child herself by the side of the road, before carrying it onward to the clinic. Mother and child were doing fine, we were told.

I escaped any such experiences. Even being out in temperatures below minus fifty was actually fine, as long as it wasn't windy, you were wrapped up warmly, and you took care not to walk too quickly. My most extreme experiences of cold were probably those times I went to Alta to spend hours freezing by a skating rink. The skaters were our biggest idols in the 1960s. Ice—or snow in the case of the cross-country skiers— was where it was all happening in those days. Norway was a winter nation—*the* winter nation. "There lies a land of eternal snow," we would sing as we paraded, flags aloft, amid flurries of snow on May 17.

But the hero of heroes was Fridtjof Nansen, who hadn't just crossed Greenland on skis—without any certainty that it was actually possible, since there were no aerial or satellite photographs then—but spent several winters in the Arctic Ocean, also something of a hit-or-miss affair. When I went cross-country skiing on the endless Finnmarksvidda, as I often did since we only had school three days a week some years, I'd daydream I was Nansen on his way across the Greenland ice. True, I wasn't hauling any baggage, I knew the weather would hold for the few hours my trip lasted, and Mom was waiting for me back home with hot cocoa, but I was Nansen all the same. Far ahead of me the west coast of Greenland awaited, along with fame and glory.

It was an almost ecstatic experience to ski across the plateau: nothing but white in all directions as far as the eye could see, just small dwarf and mountain birches dotting the white surface like tiny apostrophes, and here and there the track of a ptarmigan or hare. A view that the Danish scientist Sophus Tromholt, who studied the northern lights and lived in Kauto-keino in 1883, described as follows:

> Below a white shroud of snow the Land of the Lapps slum-bers in its winter sleep. The poor flowers, which a little while ago basked gaily in the sun, have been scattered to the winds, and only the seed remains, buried in the hard frozen earth, longing for far-away Spring, whose gentle breath shall call them into life. The thin birch copses, which used to contrib-ute their share to relieve the desolate landscape with a faint tinge of the colour of Hope, stand enveloped in Nature's common white garb, woven with the fine threads of filagree hoar frost and glittering ice crystals. The river, too, which spoke so cheeringly in the autumn, is silent, and bound in the iron grasp of King Ice.
>
> Everything slumbers after the short, bright summer's day; even the wind durst not play with the snow-white cover of Nature's couch, the very air seems to sleep. Noth-ing breaks the silence. You may wander for miles over the wastes, but never a sound, save the creak of your foot in the snow, breaks the silence either from heaven or earth.[11]

This was before snowmobiles shattered the peace of the plateau, and you could hear every tiniest sound in a radius of miles—in other words, almost pure silence. Only the noise of your skis and poles on the snow. And sunbeams coming at you from every angle, reflected by the snow crystals. Never mind that I hadn't heard of sunscreen and my face ended up covered in sun eczema: that was just part of the deal. No encounter with nature I have experienced since has lived up to Finn-marksvidda in all its winter glory. That said, I haven't skied across Greenland or Antarctica, and have resigned myself to the fact that I never will.

So my relationship to the cryosphere is largely positive, meaning that in global terms, I form part of a small minority, along with those of my fellow Norwegians who will gladly pay 70,000 Norwegian kroner (about US$8,000) for the privilege of crossing the Greenland ice on a camping trip.

Given my positive winter experiences, it was odd for me to read fairy tales and stories that described it as terrifying and dangerous, a hotbed of evil, like those of Hans Christian Andersen and C. S. Lewis. In Andersen, the wicked Snow Queen steals children and takes them with her up to her realm of frost in the north, where she travels around by reindeer, just like my neighbors the Sami reindeer herders. In the fairy tale, the boy, Kay, is kidnapped by the wicked queen and taken back to her cold palace in the north, and his friend Gerda goes after them to set him free. And in the Narnia books by Lewis, the White Witch casts a spell on Narnia, throwing it into an endless winter, in which Christmas never comes to light up a cold and dreary existence. These kinds of characters and motifs are familiar to children today through Disney films such as *Frozen*. It is clear that such stories are written in countries where people have rarely had the opportunity to experience the positive sides of winter and know only of its troublesome aspects: like snow-blocked roads and people breaking arms or legs after slipping on the ice.

The frozen world is also a popular backdrop that thriller writers from Agatha Christie to Jo Nesbø have used to sinister effect. Snow provides a setting for the most ghastly crimes and is often the murderers' accomplice, hiding their tracks when it settles on the ground like a pure, innocent carpet. Snow, ice, and frost also serve as neat metaphors for cold-blooded acts.

Yet it seems that people have a different relationship to snow and frost in Russia, which has proper winters, just as cold as those on Finnmarksvidda, and which has, moreover, been saved by winter twice in its history: first from Napoleon and later from Hitler. Both saw thousands of their soldiers freeze to death on the merciless Russian steppes. It's hardly surprising that the Russians' Grandfather Frost was the one who brought children presents, along with his beautiful grandchild, the snow maiden Snegurochka.

Some believe Hans Christian Andersen drew inspiration for the Snow Queen from the Norwegian goddess Skadi, who was actually a Jotun, but married into the family of the Aesir gods when she wedded the sea god Njord. Skadi was happiest in the cold mountains—she was, after all, the goddess of skiers—and so her marriage to the sea god fared badly. But this relationship reflects a Norse understanding of how the world originated from the encounter between cold (Niflheim, the primordial land of darkness and cold) and heat (Muspellsheim, a sea of frothing flames). Between them lay a vast, bottomless abyss, the Ginnungagap. It was here, in the meeting between fire and ice, that everything began; and it was here, too, that the world got a fresh start after Ragnarok, the twilight of the gods. But frost, Niflheim, and its children the Jotuns became the personification of evil, even though the myth acknowledges that the world wouldn't have existed without them.

The roots of the evil that fairy-tale tellers and crime writers associate with the Snow Queen's realm probably lie far back in the northern European myths of the migration period. The old Norse sagas of gods and heroes tell how the Aesir, the good guys, had to battle against the evil Jotuns—who, incidentally, weren't so evil that the Aesir didn't occasionally mate and have children with them. The fact that the Jotuns and their female counterparts, known as *gygrene* in Norwegian, came from the Kingdom of Frost is clear from their names, which originated as personifications of the frozen world.

We have Snow the Old (Snær or Snjó in Norse), who is the son of Jokul (glacier) and father to a son, Thorre (black frost), and daughters Fonn (bank of snow), Mjoll (a flurry of fine snow), and Driva (snowdrift). Perhaps all these "children" were originally supposed to be seen as different aspects of Snow, but in the myths, they took on separate roles, only fragments of which are known to us, unfortunately. In some of these, Snow the Old is the king of "Finland"—in this context, the term for northern Scandinavia, about which people knew little other than that it was cold there, with masses of snow. And that other peoples lived there—Finns, Sami, and Kvens—although nobody was quite sure who was who.

As I said, these stories now exist only in fragmentary form, so for a more coherent portrayal of how people in Norse times saw the origin of all things, we must turn to a more modern interpreter of myth, author Tor Åge Bringsværd:

> In the beginning there was Cold and Heat. On one side, Niflheim, with frost and fog. On the other, Muspellsheim, a sea of frothing flames. Between them was nothing. Just a great, bottomless abyss: Ginnungagap. Here, in this vast emptiness—midway between light and dark—all life would come

into being. In the meeting between ice and fire... the snow
began slowly to melt, and, formed by Cold but wakened to
life by Heat, a wondrous being emerged—an enormous troll.
His name was Ymir. No greater giant has ever lived.[12]

Out of the melting ice grew something else as well: the
cow Audhumla. Ymir got milk from her, and when Audhumla
licked the salty, frozen stones around her, a new wonder of
creation came about:

> The cow suddenly licked some long hair from one of the
> stones! The next day, a head and face came forth from this
> stone! And on the third day, the cow eventually managed to
> lick the whole body free. ... It was a man. He was tall and
> handsome. He was called Buri and from him all the gods are
> descended, those we call Aesir.[13]

Ymir fathered some children all by himself, from his own
sweat, and these were the origin of the "clan of the frost trolls,"
who were known as Jotuns. The relationship between Aesirs
and Jotuns was part conflict, part coexistence—the way it
often is between heat and cold. In the end, though, the Aesirs
had a showdown with the Jotuns and killed Ymir:

> The Aesirs drag the dead Ymir out into the middle of Gin-
> nungagap—the huge vacuum. They place him like a lid over
> the abyss. Here, they create the world—out of the giant's
> corpse. His blood becomes the sea; his flesh, the land. His
> bones become mountains and cliffs. His teeth and the
> crushed splinters of his bones become rocks and scree. His

hair becomes trees and grass. His brain the gods hurl high up into the air. Thus the clouds come into being. And the sky? It is the skull itself... set like a vault, a dome, above all creation. After that, the gods trapped sparks from the heat of Muspellsheim and fastened them to the heavens. There they hang to this day and sparkle.[14]

And so the world and its creatures were born from the conflict between heat and cold. Perhaps this was a myth that came naturally to those who lived where it was written down, in Iceland, a land of both ice and fire. However, it isn't so far from the newer stories modern science has given us: chunks of ice containing organic molecules strike a blazing hot Earth, causing life to come into being.

In one respect, however, Finnmarksvidda differed from the White Witch's eternal winter: winter always ended. The snow melted each spring, and even though it might last until May, it vanished quickly once the thaw had set in. The big event in spring was when the ice broke up on the great river, the Kautokeinoelva, whose name changes to the Altaelva farther downstream. Mighty forces came into play then, when the ice shattered and huge ice floes were hurled around, often far inland. Fortunately, it was safe to watch from up on the bridge, which was built to bear the brunt of it. The spring thaw was far from silent, and this, too, could make you feel kinship with Nansen, who wrote the following description of the havoc caused by pack ice in the Arctic Ocean (something his polar research vessel, the *Fram*, was fortunately built to withstand):

First you hear a sound like the thundering rumbling of an earthquake far away on the great waste; then you hear it in several places, always coming nearer and nearer. The silent ice world re-echoes with thunders; nature's giants are awakening to the battle. The ice cracks on every side of you, and begins to pile itself up; and all of a sudden you too find yourself in the midst of the struggle. There are howlings and thunderings round you; you feel the ice trembling, and hear it rumbling under your feet; there is no peace anywhere.[15]

But then it calmed. The river could once more flow down toward Alta, people could take out their riverboats, and on the hills around the river the vegetation began to peep out again after the cold winter. It wasn't long before the greenery started to appear, as the midnight sun ensured that photosynthesis—and therefore growth—continued night and day. Just a month later, you could swim in the river it was possible to drive a car across in winter. And then, after the first proper summer rain, came the invasion: billions of mosquitoes. We who grew up here became pretty much immune to their bites, but that didn't stop them from invading every cavity of your body, making it difficult to breathe. It was worst of all out on the cloudberry marshes, the fruit orchards of the plateau where I earned my summer wages. One benefit of the mosquitoes, though, was that they made autumn—and the frost—feel like liberation.

We northerners are alone in experiencing such stark changes between the seasons—from totally white to almost totally green. Farther south, the only alterations are in temperature and humidity—when the dry season is relieved by the rainy season, for example—and to some extent, the colors of the vegetation. But, with the possible exception of the moment the first rains of the monsoon come sweeping across India's brown-scorched fields, you will never experience anything so absolute as the shift from white winter to green summer up there in the north. Those of us who grew up with it yearn for the changing of the seasons; we sing songs about it, and we feel and believe this is something that all of nature experiences along with us. Or that's what we used to think before, at least, when we still spent time out in the fields and open country.

— 5 —

LIFE BENEATH
THE SNOW

Leaning toward the snowdrift, bent to the blind driving snow
A reindeer stood and sniffed the air, scraped with a cloven toe,
And all at once, as deep it dug into the frozen snows
A blue-green mossy cluster leapt toward its questing nose[16]

NORDAHL GRIEG MADE only a quick visit to Finnmarksvidda, but even so, he managed to grasp something
essential: the way living creatures—in this case a reindeer—managed to survive in this wintry land. It looks so
barren and merciless, but there is, in fact, life beneath the
snow—not least the reindeer's favorite food, reindeer lichen.

The reindeer (caribou to North Americans) was the reason
people lived in Norway's only relatively large area of permafrost. Permafrost is the ground that doesn't thaw entirely in

the summer, but only in the upper, "active" layer. A bit further below, the earth is still frozen and this makes it difficult for any vegetation other than reindeer lichen, heather, and dwarf birch to thrive. Originally, people followed the reindeer herds as they headed north after the last ice age in order to hunt them—just as they had during the ice age, although farther south in Europe in those days. Traces of the hunt are visible in the wealth of rock paintings, especially near the Pyrenees, where reindeer and people lived during the ice age.

Several hundred years ago the Sami people (formerly known as Lapps or Laplanders), thought by many to be the first humans to settle here in the north, began to domesticate reindeer instead of hunting them. In other words, they followed the reindeer on their annual migrations: from the plateau, where they fed on reindeer lichen in the wintertime, to the grassy pastures near the coast in spring, and back to the plateau again in autumn. Gradually, the reindeer got used to the humans, although they never became totally tame. This is how a lifestyle that anthropologists call "semi-nomadic" came about, in which humans follow the animals' seasonal migrations between two fixed grazing areas—one in winter and one in summer. The pattern of migration is now so established that Sami reindeer herders have permanent homes for both winter and summer use but spend several weeks of each spring and autumn living, literally, on the move. Particularly in spring, when you never know quite when the snow and ice on the lakes and rivers will melt, it can be pretty demanding, and sometimes dangerous, especially for the reindeer. For many reindeer, the migration also involves swimming across a sound in ice-cold water, which is generally the most critical point in the journey.

Contrary to what the schoolbooks used to tell us, however, only a minority of Sami people, those known as *flyttsamer*, live this way. Most live "normal" lives as farmers and fishermen ("sea Sami"), and as nurses, machine operators, teachers, newspaper editors, job seekers, pensioners, and clothes designers ("city Sami"). But the nomadic group are the "prototypical" Sami who, in many ways, keep the traditional Sami cultural traits alive. Most important of all these is reindeer herding and the related cultural practices: the *lavvu* (the tent used during the migration); the clothes and tools they carry with them, which are optimally designed for this purpose and are generally made of reindeer hide, antler, or bone; the traditional reindeer races now held at Easter; the meals of bone marrow; the lasso throwing. We can see many examples of reindeer herding as a motif in the work of the best-known Sami artists: John Savio, Iver Jåks, and Nils-Aslak Valkeapää.

For the Sami people who live with and from the reindeer and move with them twice a year, reindeer have been their alpha and omega. And I really do mean the whole alphabet. Reindeer have given them everything they need. Clothes, for a start: in winter they used *skaller*, footwear made of reindeer hide, fur side out. And in summer, they used *komager*, also made of reindeer hide, but without fur—or hide boots known as *bieksoer*, which were also popular among Norway's "alternative" people in the 1970s. When it got really cold outside, people would wear a *pesk*, a kind of fur coat. Also tools such as needles and combs used to be made of reindeer antler and bone.

The diet was dominated by reindeer meat, too, generally dried and dipped in boiled black coffee to make it easier to chew. Party food was boiled bone marrow or a reindeer soup known as *bidos*. Reindeer meat was also a source of income. Previously, the reindeer owners slaughtered their animals themselves, but nowadays they deliver them to the abattoir, other than those the herders retain for their own use.

Last but not least, reindeer were also the Sami's draft animals, harnessed to sleighs or sleds known as *pulks*—when there was snow, that is. In summertime, the reindeer were pretty much free, although people might sometimes use them as pack animals. In winter, the reindeer offered the ideal means of transport, reliant on neither roads nor bridges. The whole plateau lay open to those able to use them.

Snow was also the reindeer's element in another way. Beneath the snow, properly insulated and therefore not frozen regardless of how far below zero temperatures lay, the reindeer could dig down to their main winter food: reindeer lichen. They were, in fact, dependent on the snow and its unique insulating properties: if not for the snow, the reindeer lichen would have frozen and the reindeer would have starved—as can happen in winters where snow is scarce. Then a layer of ice forms over the lichen and the reindeer cannot reach it.

Frozen reindeer lichen also brought a TV series to a temporary halt. When the Norwegian state broadcaster, NRK, was about to launch a "slow TV" series following the spring migration of a family of *flyttsamer* and their reindeer herd, the reindeer didn't want to leave. The reason was that the snow had first melted in a period of mild weather and then frozen again into ice. This made it difficult for the reindeer to find reindeer lichen and they responded by postponing their migration, staying where they were until the ice began to thaw again and the reindeer lichen became accessible. The leader does were the ones that decided when the migration should start, and the TV people just had to wait patiently.

For reindeer, snow is a gift from the gods, and that has also been true for the humans on the plateau. The snow has made it easy to travel, independently of roads, using either reindeer or skis. The Sami people were probably Norway's first skiers, as Sami director Nils Gaup suggested in his film *Pathfinder*: the Sami can travel on skis, whereas the Tjudes, a marauding people, have to struggle through the snow on foot. Rock carvings of skiers dating back several thousands of years have been found in Alstadhaug and Alta—the images that inspired the distinctive icons of skiers used at the 1994 Winter Olympics in Lillehammer.

The Sami people's skiing skills were also well known to Fridtjof Nansen, who took two Sami men with him when he skied across Greenland: Ole Nilsen Ravna and Samuel Balto. The same went for the Finnish-Swedish explorer Adolf Erik Nordenskiöld when he crossed the Greenland ice in 1883, as well as Carsten E. Borchgrevink when he went to the Antarctic in 1900. The two Sami men he took with him, Per Savio and Ole Must from Finnmark, were the first people to spend a night on that continent. And they didn't do this because they were adventure seekers, by the way, but because they were abandoned there temporarily after bad weather forced the ship to put away from the shore.

That the Sami people had mastered snow, whether traveling by ski, reindeer, or snowmobile—the option most of them use now—was something I learned as a child on the Finn-marksvidda. I had trouble keeping up even with those who used traditional equipment—skis with simple leather straps into which they slipped their *skaller* (today, of course, they use modern gear). What I didn't know then, but have since learned, is that the Sami people also have an incredibly rich vocabulary for all things snow related. Even their language is adapted to life in the Kingdom of Frost.

— 6 —

MORE THAN
A HUNDRED WORDS
FOR SNOW

M**ANY PEOPLE HAVE** probably heard about the discussion of whether Inuit really do have a hundred words for snow. Anthropologists and linguists have cast doubt on the idea, regarding it as a romantic perception of Indigenous people's traditional knowledge. I don't know the truth about the Inuit vocabulary for snow, but I reckon that people living in such close contact with snow and ice actually do need a detailed vocabulary for it. What I do know for sure, though, is that there is one people with many *more* than a hundred words for snow: the people I grew up with on the Finnmarksvidda.

Language experts like Nils Jernsletten,[17] professor of languages; Ole Henrik Magga,[18] better known as an Indigenous politician and the first president of the Sami parliament; and associate professor Inger Marie Gaup Eira[19] have collected and analyzed Sami snow terminology. According to Magga, the North Sami language (the most widespread form of Sami) contains 175 to 180 root words for "snow" and "ice." Once you factor in derivations, inflections, and variants of these roots— for example, the noun *njeadgga* means "drifting snow," while the verb *njeadgat* means "to drift" and the adjective *njeadgi* refers to a type of weather involving snowdrifts—the number of words for snow and ice adds up to around a thousand.

This finding also applies in other Sami languages, such as Lule Sami (named for the Lule river valley in Sweden). The reindeer herder Johan Rassa was born in a *lavvu* in the mountains of northern Sweden in 1921 and was one of the last bearers of a rich tradition of knowledge about snow in the Sápmi region, known as Sábme in Lule Sami. The author Yngve Ryd spent five winters speaking to Rassa about snow and the Lule Sami terminology. His book, based on Rassa's knowledge, explains more than three hundred words related to snow and ice, together with the context in which they were used. This rich terminology was one aspect of the reindeer-herding Sami people's adaptation to snow, which dominated the landscape for seven to eight months of the year: "A highly detailed knowledge of snow and ice previously went hand in hand with the business of reindeer herding. Weather, wind, and snow used to be everyday topics of conversation."[20]

One reason why knowledge of snow was so vital, as reflected in this compendious vocabulary, was that snow—how *much* snow, what *kind* of snow, and so on—could drastically alter living conditions for humans and animals: "People's reliance on snow and ice shifts between extremes, and this may have contributed to the rich vocabulary. It might snow in such a way that it is a struggle to move even a few hundred meters, and yet snow can also provide such good skiing conditions that one can easily whizz along for a couple of dozen kilometers."[21] People used so many words about snow and snow conditions because they were talking about snow from different points of view, in different situations. The topic was never snow as an "objective," physical entity but snow as one approached it, used it, and had to adapt to it. It was how snow fell in autumn, in winter; how it remained on the ground from winter through to spring. It was how snow behaved in relation to people and animals, restricting their movements or grazing possibilities—a vocabulary about snow that grew through practice.

There are many reasons why snow conditions are extremely important for reindeer and their herders. First of all, as mentioned earlier, for much of the year the reindeer must find their food—reindeer lichen—by digging beneath the snow. So it is vital for them to do this where it is possible to dig through the snow, which is difficult if it has previously thawed and then refrozen. And while snow is a unique insulating material when it is light, owing to the air inside and between the snow crystals, both ice and snow that is more akin to ice are a different matter entirely. The snow also determines how reindeer and people can travel. In some kinds of snow, movement can be almost impossible, while in others it is child's play. And this doesn't just apply to deer and people: predators such as wolves and wolverines also rely on snow conditions—a fact that reindeer

yardstick, you can speak of *gámamuohta*, which means shoe snow. That's when there is not much snow and it just about reaches the top of your shoes. *Vuottamuohta* is snow that reaches your shoe bands. *Vargga buolvvaj* is snow that comes almost to your knees. *Buolvvamuohta* is snow at knee height, while *badárádjmuohta* is snow that reaches your buttocks. Masses of snow are called *giedavuolmuohta*: snow that comes up to your armpits. But the most important measure for people is *tjibbemuohta*, snow up to your shins, the same snow depth as *doavgge*. It is a special event when *tjibbemuohta* arrives: "Now it has snowed *tjibbemuohta*. We can't walk anymore; it's time to get out the skis."[24]

Snow terminology is different when reindeer are the yardstick. The most important word is *tjievttjemuohta*: this means snow that reaches to the knees of the reindeer's hind legs. It's a bit deeper than *doavgge*, toward 16 inches. With *tjievttjemuohta*, the reindeer's hooves can touch the ground, so they can still walk and travel with relative ease. There is also a special word—*doalli*—for when the snow covers an old track, where the reindeer can gain a better foothold—and they seem to have a special sense that helps them to find their way to such places.

There are different words for falling snow: single snowflakes are called *muohtatjalme*, which translates literally as "snow eyes." Big soft snowflakes are called *tsihtsebelaga*, while the driest, lightest snow that can fall in winter is *habllek*. The flakes are so big and weightless, sometimes they almost refuse to fall. While humans can cope with this snow, it is dangerous for animals, which can actually be suffocated by it. This is why it was common to hunt fox when there was *habllek*. A little new snow, say 1 or 2 inches deep, that settles on top of previous snowfall is called *vahtsa*. *Loahtte* is a heavy snowfall of 8 inches or more. *Larkkat* occurs when there has been a heavy fall of dry snow that has stopped quite abruptly. The worst precipitation is sleet, *slabttse*. This is problematic because it can't be brushed off like dry snow but sticks to clothes and other places, making people wet.

A lot also happens to snow when it has settled, and then its name changes. Thin and slightly icy snow on the ground is called *skártta*. *Tjalssa* is soft snow that is trampled solid and freezes to the earth. Nearest the ground, the snow gradually turns into *sänásj*, large, coarse, icy grains that look like coarse-ground salt. It can also become *tsievve*, hard snow where the reindeer don't dig. This kind of snow can almost bear the weight of a person without skis and the reindeer float on top of it, whereas *åbådahka*, or simply *åbåt*, is thick winter snow that is very soft and loose. *Åbåt* creates extremely difficult conditions. In the old days, people hunted wolves when there was *åbåt*, because then the wolves would become exhausted. *Dáhapádahka* means that conditions are so poor it is impossible to travel at all, while *siebla* is snow that has thawed and is wet all the way through to the ground, a typical spring phenomenon. *Siebla* cannot bear any weight: the skis sink straight through it. When *siebla* freezes it becomes *tjarvva*, a proper snow crust.

But the navigability of terrain is not the only important thing. For hunters—and hunting was (and remains) an important part of the mountain Sami livelihood—it is important for skis to move silently across the snow. So conditions that allow skis to glide quietly and softly, what we call "silk conditions" in Norwegian, are known in Lule Sami as *linádahka*.

The mountain Sami also had a rich vocabulary for ice, because in their world it was vitally important to be able to move over streams, rivers, and other water, so people needed to have words that told them whether the ice could bear the weight of people and animals. The first very thin ice to form on the lakes in autumn is called *gabdda*. It is barely one twenty-fifth of an inch thick. *Álmasjjiegŋa* is "people ice," which can bear the weight of a person on foot, while *hässtajiegŋa* is ice that can bear a horse.

As with so much other traditional knowledge, this Sami snow terminology is disappearing. Fewer people are involved in reindeer husbandry, and since reindeer herders use snow-mobiles rather than draft reindeer these days, perhaps they think they don't need this "old-fashioned" knowledge. But the many snowmobile accidents, not least those where avalanches are triggered, may well suggest that modern-day reindeer herders could also benefit from a bit of knowledge of snow. Would it help to have a special word for ice that can bear the weight of a snowmobile? Perhaps the technicians waxing skis for the Norwegian cross-country team would also find it helpful to have a course in Sami snow terminology to avoid waxing blunders.

But this is about a lot more than an advanced vocabulary, rich in tradition, that is on the point of extinction. What we are now seeing vanish is also a *lifeworld*—the world of the Snow Queen—to which this language belongs and which it describes.

TRACES OF ICE: THE DISCOVERY OF OUR FROZEN PAST

NEVER LEARNED THE three hundred Sami words for snow. I had to leave Finnmarksvidda early in my teenage years to go to school and I never went back. My first stop was the Arctic Ocean town of Tromsø, where I experienced a slightly different side of the cryosphere: severe snowstorms and endless snowfall. Some winters, so much snow fell that you just had to give up trying to keep the path clear and simply dig a tunnel to your door instead. It could stay like that until late April. On the other hand, Tromsø had fantastic skiing possibilities, offering the unbeatable combination of skiing and sea views. Here you got to experience the cryosphere at its best and worst. Some-times there could be a bit much of both, but if you loved snow, it was fine. If you didn't, you moved away.

Eventually, I ended up in western Norway, where snow and ice were things you only experienced once in a while, mostly as unforeseen problems—like suddenly waking up to find ice and snow on the roads, which came as just as much of a sur-prise every single time. Fortunately, this happened only a few days a year, so instead of bothering to change to winter tires, people tended to leave their cars at home. Snow and ice were mostly irrelevant there on the west coast. They were curios-ities, of interest mainly to skiing enthusiasts—only a small minority there—not to mention tourists, who came in their thousands on cruise ships to visit Nordfjord and see the Briks-dal Glacier close up.

Since I was neither a skier nor a tourist, snow and ice didn't concern me. I thought I was done with the Kingdom of Frost and didn't even realize that I was still wandering around in the midst of it. I didn't see that the whole landscape in Norway had been shaped by the ice. The distinctive, world-famous fjords and valleys, the huge erratic boulders you could see in the most peculiar places—all of this was the work of the ice. And I was far from the only person who was blind to it. Peo-ple had walked around here for hundreds, thousands of years without realizing it. They just took for granted that the land-scape was the way it was and didn't ask why.

Indeed, it wasn't until the mid-1800s that people learned that the glaciers had carved out these U-shaped valleys, leav-ing fertile earth in the valley bed. People had no idea that there had once been glaciers here. There was nothing about it in the Bible, which was the most important history book until the 1800s. Most people then believed the world had existed for only 6,000 years. And the term "ice age" was totally unknown.

It took a Danish immigrant to discover that there was something strange about this, something that demanded an explanation. The huge stones that often stood balanced on top of a hill couldn't have ended up there all by themselves. Stones don't roll uphill. And the huge, continuous ridges of rock and gravel down in the valley: who or what had created them?

It must have been questions like these that Jens Esmark (1763–1839) pondered during his many journeys around Norway. Esmark was a trained "mineralogist," as geologists were called in the late 1700s when he arrived in Norway, initially to work in Kongsberg. At that time, it was an important mining town and also had an institution called the Bergverkseminar, known in English as the Kongsberg School of Mines, which is where Esmark taught. After the company in Kongsberg went bankrupt in 1805, Esmark moved to nearby Christiania, modern-day Oslo. There, he became the first professor of mineralogy at Norway's first university, which opened in 1811.[25]

While he was living in Kongsberg, Esmark had already managed to travel around and familiarize himself with the Norwegian mountain regions. He was probably the first person to climb mountains like Snøhetta and Gaustatoppen, and he also undertook height measurements of both using barometers, which measure air pressure. This enabled him to prove—to many people's surprise—that Gaustatoppen, hitherto considered the "roof of Norway," was not Norway's highest mountain but was actually lower than Snøhetta.

On his journeys to map the geology of the Norwegian landscape, Esmark traveled to places such as Lysefjorden in western Norway. There, at the end of the lake called Haukali-vatnet, lies a terminal moraine, which we now know was deposited by a glacier. Before Esmark proposed this theory in 1823, nobody had guessed that this formation was created by a glacier that had long since disappeared.

Eventually, he discovered similar traces in many places and wrote an article in which he set out his theory that there must once have been glaciers across the whole of Scandinavia, which had created the characteristic formations for which the Norwegian landscape is so famous.

Esmark also had his ideas published in English in 1826 but failed to attract much attention. However, the renowned professor Robert Jameson, who was at one time Darwin's teacher in Edinburgh, gave lectures about Esmark's ideas and it is quite possible that they spread further from there. One of Jameson's contacts was the Swiss natural scientist Louis Agassiz (1807-1873). We have no proof that Agassiz became aware of Esmark's ice age theory through Jameson, but it is highly unlikely that Jameson would not have told Agassiz about it; Jameson was Agassiz's English publisher.

At any rate, it was Agassiz who won all the glory for introducing the theory of the ice age, while Esmark vanished into oblivion. That is often the way with science: it isn't necessarily the person who came up with an idea who wins glory for it but the person who first manages to disseminate it to a wider public—as Agassiz did, initially at a famous lecture in Neuchâtel in 1837. The theory of the ice age made waves in scientific circles and was met with skepticism. Even renowned scientists, including the German naturalist Alexander von Humboldt (who had been Agassiz's teacher), slaughtered the idea. But the data supporting the theory were impossible to explain any other way.[26]

Being from Switzerland and very familiar with the Alps, Agassiz had noticed some of the same phenomena Esmark had seen in Norway: for example, the erratic boulders and the moraine ridges. Agassiz had seen how modern-day glaciers could carry such rocks along with them, and how they formed moraines from stones and gravel. He also noticed the striations on the surfaces of the rock, which all went in one and the same direction, as if somebody—or something—had scraped the rock with some vast instrument.

Eventually, Agassiz was able to picture how the ice had covered a much larger area than the small glaciers that still remained up in the mountains. He latched on to a term one of his friends, a half-crazy German botanist called Karl Schimper, had used in a poem in 1837: ice age. Agassiz gathered more proof and in 1840, he was able to publish his revolutionary theory: that large swaths of Europe, and perhaps other parts of the world, had once been covered in ice, and that this could explain many of the landscape formations. As he put it: "In my opinion, the only way to account for all these facts and relate them to known geological phenomena is to assume that . . . the Earth was covered by a large ice sheet that buried the Siberian mammoths and reached as far toward the south as did the phenomenon of erratic boulders."[27]

Agassiz was here referring to the mammoths that had created a furor when well-preserved specimens were discovered in the Siberian tundra. The discovery of these mighty frozen creatures helped give his theory credibility. At first, it is true, people thought they were elephants, washed north to Siberia by the flood described in the Old Testament. But the French natural scientist Georges Cuvier (1769–1832) proved mammoths were a separate species, specially adapted to life in the cold Arctic.

In spite of severe opposition—this contradicted the Bible, after all, and predated Darwin's theory of evolution—Agassiz's theory was eventually accepted. And now, suddenly, the traces were easy to see and people found them almost everywhere: moraines, erratic boulders, striations, U-shaped valleys, or U-shaped fjords, as in Norway. The same types of traces also appeared in other parts of the world, such as North America and New Zealand. Old stories about how the glaciers had grown or shrunk, previously dismissed as tall tales, were looked at again. Agassiz himself moved to the United States, where he garnered considerable recognition as a scientist and became a professor at Harvard. In his new homeland, he saw many signs that North America had also had an ice age. The great lake that formed when these glaciers melted was named Lake Agassiz.

Aided by numerous ingenious methods, scientists have since discovered that there hasn't been just one ice age, but many. In the past 800 million years, in particular, Earth has frozen and thawed, frozen and thawed in a cyclical dance. In the past 800,000 years alone, there have been at least *nine* ice age "episodes." The first thorough documentation of this was in the 1960s and '70s, with British scientist Nicholas Shackleton's analysis of sediments from the ocean bed, together with Danish paleoclimatologist Willi Dansgaard's ice-core drilling in Greenland. Both made use of the fact that the composition of oxygen isotopes (variants of oxygen with different atomic weights) changes in step with temperature and sea level. On that basis, it was possible to produce time series that showed how the climate had fluctuated in the last ice age periods. Among others, they showed that Earth has spent most of the past million years in ice ages and that it has been much colder than it is in our times. Even though solar radiation has increased, Earth has basically become colder. According to this pattern, we should now actually be on our way back to a new ice age. But because of what we are currently doing to the atmosphere, it is far from certain this will happen.

WHAT CAUSES ICE AGES?

WHY ALL THESE fluctuations, these shifts between cold and hot periods? Isn't the temperature on Earth primarily determined by solar radiation? And in the 4.5 billion years in which Earth has orbited the sun, that radiation has, in fact, increased by as much as 30 percent. Surely this constant increase in heat from the sun shouldn't lead to ice ages?

There has been a lot of speculation about and research into the possible causes of these climate fluctuations—an especially burning question now, in light of global warming. Part of the answer, scientists now agree, was discovered by the Serbian engineer Milutin Milanković (1879–1958) when he managed to demonstrate links between temperature fluctuations and cyclical changes in the Earth's movements. Two elements that come into play here are the Earth's distance from the sun and its axial tilt.

We know that the changing seasons are caused by the fact that the Earth's axis of rotation is not perpendicular to the sun. It is slightly tilted, by between 22.1 and 24.5 degrees on average, so that the North Pole points away from the sun when it is winter in the north (the sun does not rise north of the polar circle in the so-called polar night). Likewise, it points toward the sun when it is summer in the north, so that for a time, the sun does not set north of the polar circle ("midnight sun"). When it is winter in the north, it is summer in the south, and vice versa. At the equator, there is no such difference between summer and winter.

But Earth's axial tilt—and with it the differences in the seasons—also varies over longer time spans. It is believed that this, together with variations in Earth's orbit around the sun, is the reason why Earth's climate has moved in and out of ice ages, interrupted by shorter interglacial periods. Milanković discovered that these changes followed three cycles, which might seem a bit confusing, but the point is that there are three types of changes in the Earth's rotation, which overlap one another.

The first cycle results from the fact that the elliptical shape of the Earth's orbit around the sun varies in "eccentricity": sometimes it is more like a circle, other times more like a stretched-out ellipse. This cycle has a periodicity of 100,000 *and* 400,000 years. The second cycle, which has a duration of 41,000 years, is caused by the fact that the Earth's axial tilt (how far it leans) varies between 22.1 degrees and 24.5 degrees. But this second cycle itself varies, in a *third* cycle which varies over two periods of 19,000 and 23,000 years, and which is, in a way, comparable to a spinning top that points this way and that when it is turning slowly. Whereas Earth's North Pole currently points toward the North Star, in 11,000 years' time, it will be pointing toward Vega.

These variations create small but important differences in temperature in the north and south, which scientists believe trigger ice ages (the time patterns concur), although the differences in themselves are not enough to cause climate changes as dramatic as those Earth has in fact experienced. Other effects have also come into play, exacerbating the original differences. Gradually, it has become clear that changing conditions on the Earth's surface and in its atmosphere also influence the climate. Earth is not a huge, dead chunk of rock: it has a hot core that periodically spews up substances; it has shifting continents; most of its surface is covered in water; and it has air, an atmosphere. All of these are moving systems, which affect one another and the climate through an interaction that is so complex and full of feedback mechanisms that scientists have not yet understood precisely how it works.

One effect caused by these systems is the much-discussed greenhouse effect (see "A Closer Look: The Greenhouse Effect," chapter 3). Studies of small air pockets left in rocks and ice show that the concentration of greenhouse gases has varied dramatically over Earth's history. Today, we think the most important source of the greenhouse effect is carbon burning, but this is just a fraction of the truth, and it was not the case in the first few billions of years at any rate. Back then these gases came from the Earth's core, through volcanoes and similar "vents." Since volcanic activity could be high in Earth's earliest phases, the concentration of greenhouse gases was also high. That is probably why there were no ice ages in the first 1.6 billion years of the Earth's history, even though the sun's radiation was weaker then and Earth "ought" to have been colder. So the greenhouse effect is nothing new; indeed, it has at times been much more severe than it is now.

Another factor of decisive importance is the movement of Earth's continents: continental drift (a phenomenon explained by the theory of plate tectonics). After our planet had cooled down a bit, solid landmasses formed above the blazing hot, more fluid core. But because the heat of the Earth's core prevented it from solidifying entirely, these landmasses continued to shift about. They still do, although fortunately this happens slowly. At one point, continental drift left most of the Earth's landmasses close to the equator. The point where the Earth felt most heat from the sun was therefore mostly land, which reflects sunlight better than the heat-absorbing sea. At this time, there was no vegetation to damp the reflection either. Since the reflection of solar energy (what we call the albedo effect) increased, this caused the Earth's temperature to fall further.

Now, yet another factor came into play: *life*. And although initially we are only talking about simple microorganisms, the total biochemical effect of hordes of these may have been a major driver of—or brake on—the greenhouse effect. However, the extent to which these organisms export (emit) or import (absorb) carbon depends on what kind of organisms we are dealing with and on the environment around them.

But all this took time. More than a billion years would pass before precipitation could fall as snow and not rain. Eventually it also became possible for the snow to remain on the ground and become ice. This happened 2.9 billion years ago, as we can tell from the striations the ice made in the rock, which can be dated using radioactive isotopes. The same sorts of traces also tell us that a series of glaciations appeared half a billion years later (i.e., 2.4 to 2.3 billion years ago). There is much to suggest that at that time (and perhaps also 2.9 billion years ago) the Earth was entirely covered in ice and snow, in a phenomenon known as "snowball Earth."[28]

It was a Caltech scientist, Joe Kirschvink, who introduced this term together with his snowball Earth theory in the 1990s. The theory initially caused controversy, not least because many people believed that if Earth had once frozen in this way, it would never have been able to thaw again. The albedo effect—the fact that white reflects almost all the solar energy— would ensure that. However, the geological and chemical proof that this kind of glaciation had, in fact, occurred became compelling. And data models showed how it might have happened: once a large area had become white, the albedo effect would set in motion positive feedback, a self-reinforcing process.

But how, then, could the white melt again, returning Earth to "normal"? Well, there are feedback effects that take care of that too, as British scientists Tim Lenton and Andrew Watson have shown in their book *Revolutions That Made the Earth*. When everything was covered by ice and snow, natural carbon dioxide uptake by the ocean and the Earth came to a standstill. And since volcanoes and other emissions from the core of the Earth continued to spew carbon dioxide out into the atmosphere, this led—after several million years—to the buildup of a greenhouse effect that warmed the planet, causing the snow and ice to start melting. It also helped that the volcanoes rained ash and other dark particles onto the snow. And once the melt had begun, it unleashed positive feedback in the other direction: when the sea opened up, it absorbed the heat of the sun, thereby accelerating the melting of the ice, leading to greater absorption of the sun's heat, and so on.[29]

Another factor that affected the climate was that life had begun to develop on the planet, in the ocean, before the first snowball Earth episode. It is, of course, difficult to say just when the first living organisms came into existence on Earth. But traces from Greenland show that life already existed there 3.5 billion years ago. There were simple, single-celled organisms, prokaryotes, that could actually survive the freeze—in other words, they could thaw out again after a snowball Earth period that had lasted several million years and carry on as if nothing had happened. Some of these, cyanobacteria, began to produce oxygen, which was then scarce and originally an environmental poison, but which eventually altered the composition of the atmosphere. (One way of finding out whether there is life on a planet is to check whether there is oxygen—or, in practice, ozone—in the atmosphere.) This change, the increasing volume of oxygen in the atmosphere and an equivalent decline in carbon dioxide levels, also contributed to changes in the climate, making it grow cold enough for the first glaciation. Many complex chemical and geophysical processes were involved, over a long period of time, and nobody has managed to fully explain the entire sequence of events. But it seems that the Earth has indeed had feedback mechanisms capable of thawing it out again after the repeated snowball episodes, and that the first shoots of life actually managed to survive these events.

After two or three freezes, around 2.4 to 2.3 billion years ago, the Earth entered into what might be seen as a rather boring period, without any major, visible events for almost 2 billion years. But something important was happening all the same: the oxygen content in the atmosphere built up slowly. Oxygen—and the energy it could offer—was absolutely vital in order for life to take its next big step, to multicellular organisms. Before this could happen, the last two known snowball episodes occurred, 710 and 640 million years ago. The glaciations weren't over—quite the contrary. But they were never again total: they were restricted to the areas around and somewhat beyond the poles. For us, this was fortunate, since the multicellular organisms that now appeared would not have been able to survive any snowball episodes. Evolution would have had to start almost from scratch after every glaciation. But it may also be that life itself had now altered the atmosphere and the chemistry so much that these total glaciations could no longer occur; that, in a way, life itself had created an insurance against them.

Now the stage was finally set for the white caps to embark on their dance to and fro: the ice ages came and went but no longer threatened to cover the entire planet. It was like learning a new, slightly lively dance: at first, you may do the steps wrong and perhaps you may even fall over. But after a while, you get better control and learn to master the movements. And that allows you to dance more intensely, with faster and faster movements.

Of course, there is no master plan or "cosmic choreography" behind this. Nonetheless, it is a fact that in the past hundred million years, the Earth has engaged in an ever more vigorous dance, with major fluctuations between cold and hot periods, in which the white caps of ice and snow have expanded and then retreated.

And thus the Earth's climate has continued to swing, thanks to a combination of volcanic activity, the displacement of landmasses through continental drift, the resultant change in ocean currents and winds, and the new element (oxygen) introduced by new life forms—from microorganisms (on land and sea) and vegetation, which absorbs carbon dioxide. All these factors have worked together in a complex interaction which we have not yet fully grasped.

What we know most about, naturally enough, is the last great time of ice ages. This started some 50 million years ago and was most probably set in motion by reduced carbon dioxide levels in the atmosphere. There may have been several causes: perhaps there was a period of lesser volcanic activity. The Earth's core had also cooled sufficiently for the landmasses, the continents, to begin to settle and find their place—where they are today. Once the continents were where they were "supposed" to be, the routes of the planet's most important ocean and wind systems were also established. And since sea and wind are what distribute the energy the Earth gets from the sun across the planet and toward the poles, the arrangement of the continents would have vital significance for the climate and the cryosphere.

We know that it was in this time that the Indian subcontinent collided with the Eurasian, and that this led to the formation of the mountains of the Himalayas and the adjacent regions. That in turn created one of the planet's most important weather systems: the monsoon, which sweeps in every year from the Indian Ocean, providing precipitation and humidity for millions of people in Southeast Asia. It also set in motion the weathering of the bedrock, which caused newly formed rock to react chemically with the atmosphere, absorbing large quantities of carbon dioxide. As a result, the carbon dioxide content of the atmosphere—and therefore the greenhouse effect—diminished, and this in itself led to cooling. This cooling reached a threshold around 34 million years ago, when the Antarctic began to gain ice cover, which unleashed a powerful feedback process (because snow and ice reflect sunlight and therefore heat). This process really took off when the sea area around Antarctica was opened up around 25 million years ago, after continental drift opened the passages between Antarctica and South America, and toward Australia. Antarctica was left isolated and a cold ocean current now circulated around this chilliest of continents, ensuring that it became even colder. The time when animals—even dinosaurs—ran around the forests of Antarctica was now definitely gone.

Scientists calculate that East Antarctica had permanent ice cover from around 15 million years ago, and West Antarctica some millions of years later. Gradually, the ice cover grew to an average thickness of more than a mile. In the north, in the Arctic, this came a bit later. Glaciers began to form on Greenland around 10 million years ago. But the ice here and elsewhere in the Arctic has fluctuated in tandem with the cycles of the ice ages, which have been pretty turbulent over the past 2.5 million years. Ice ages, in which large stretches of the northern hemisphere have been frozen, have been succeeded by shorter warm periods, when almost all the ice—other than in the Antarctic and on Greenland—has melted away.

The very last ice age started around 116,000 years ago. Glaciers began to develop in Canada and Scandinavia. Then the ice spread out, in fits and starts, across North America, Europe, and parts of Asia, South America, and New Zealand. It peaked just 21,000 years ago, before a new interglacial period—the era we live in—took over, with a couple of interruptions. At that time, the whole of northern Europe and large swaths of North America were covered in ice, and our forefathers and foremothers had sought refuge farther south in Europe, in particular down by the Pyrenees. The ice covered Scandinavia, most of modern-day Great Britain, as well as parts of north Germany and the other countries on the Baltic Sea. South of the ice lay the tundra, inhospitable to humans but offering good conditions for animals like reindeer and mammoths. These, in turn, were hunted by big-game hunters, who had developed effective hunting techniques. Around 17,000 years ago, the ice age began to abate and a rapid but slightly uneven warming followed. The glaciers retreated, the tree line moved northward, and animal life followed it. Some big-game hunters followed reindeer herds north, while some switched to hunting other animals like deer and harvesting the fruits of the forest, such as nuts.

And then, around 12,000 or 13,000 years ago, after a period of rapid warming, the ice age climate suddenly returned, causing temperatures to fall a full 18 degrees Fahrenheit over the course of a few years in some places. A thousand years after that, it rebounded just as quickly. Today's temperature swings look pretty puny in comparison with such fluctuations.

Because we live our lives on a timescale that is much shorter than the geological timescale and also have a poor

— 8 —

PARADISE

T HERE ARE SOME places on Earth it seems natural to call a "paradise"—places we would happily call home if we had the opportunity and the money, or would at least choose as a holiday destination. Places where the sun shines and the temperature is pleasant even in the sea. But sun and heat are not enough: in order to merit the name of paradise, the landscape must be lush, overflowing with "milk and honey" and, preferably, with all kinds of different fruits. Including grapes.

California is one such place. Few regions have had so many pop songs written in their honor: "California Dreamin'," "Back to California," "California Calling," "California, Here I Come," "California Paradise," "California Nights," "California Rose," "California Song," "California Soul," "California Sun," "It Never Rains in Southern California." There's even a song called "We Don't Need Another Song about California," not to mention countless songs called simply "California." Wikipedia lists over a hundred of them with that title, including one by Nobel Prize winner Bob Dylan.

Several of these songs run through my head as I drive from beautiful San Francisco Bay, cross a small mountain range, and see the Central Valley, one of the world's most productive agricultural regions, spreading out before me. Because what makes California a paradise is not just the dream of the good life on the beach that the Beach Boys and others sang about. It is also fruitfulness. Mile after mile of citrus fruit and almonds, and a bit farther north in Napa and Sonoma, endless rows of grapevines. It is easy to forget that it wasn't always this way, and that a lot of work lies behind this, as Steinbeck described in *The Grapes of Wrath*.

California is, perhaps, in a unique position and has been since the 1849 gold rush. But other areas of the world offer much of the same, also redolent of the good life and overflowing with wine, citrus fruit, and other such delights. Many of them are around the Mediterranean, and those that are not— California, the Western Cape (South Africa), New South Wales (Australia), and others—generally share some common features with Provence, Liguria, and Costa Blanca.

First and foremost, they enjoy a typical Mediterranean climate. In other words, the temperature is pleasant all year round, but there are typically some wet months and other periods that are drier and hotter. That suits us tourists fine: we're only there to enjoy ourselves and can do that in the sunny months. But the dry periods can be a problem for the people working the land, especially if it gets seriously hot, with temperatures topping 100 degrees Fahrenheit.

In order for a place to qualify as a paradise, it must therefore also have access to water in the dry months that form part of the Mediterranean climate. In some cases, this can come from far-traveled rivers, but it is more usual for the typical paradise to have some high mountains nearby, where there is either heavy snowfall in winter or glaciers that can serve as water towers. The meltwater from such sources helps keep the paradise fruitful even in the driest and hottest months of the year.

This is how it is in the areas south of the Alps, in northern Italy and southeast France (Provence). This is how it is in the valleys east and west of the Andes, in the vineyards of Argentina and Chile. This is how it is in Andalusia and southeast Australia. Water comes from the mountains, which are generally covered with snow in winter (New South Wales has its own Snowy Mountains and the name of Andalusia's and California's Sierra Nevada has the same meaning) and perhaps have glaciers that people aren't aware of, because they are covered in gravel, as in the central Andes.

These are the kinds of things that transform such areas into an earthly paradise. The same goes for California. True, some rain may come at times, but for most of the year, the climate is extremely pleasant without being too dry. The snow that melts in the Sierra Nevada (and the glaciers that are still to be found there, partly hidden) ensures that water always comes: to the vineyards and almond trees, to the lawns and golf courses, and to the rivers where the salmon still return in spawning season.

Without snow, without glaciers, California and the other regions I have mentioned would be anything but paradise. Then the hot, dry months would be unpleasant for plants, animals, and people. It is precisely because of the frozen world, because water is stored in the mountains as snow and ice that can be released gradually in the hot season, that those hot months appear so pleasant, so heavenly. And it is because of the meltwater that these regions in particular are among the Earth's most fertile. That's something to think about as you sit enjoying a Californian Zinfandel accompanied by cheese from Provence and olives from Andalusia. They taste of the Mediterranean, of sun and heat. But they wouldn't have been possible without snow and ice. The cryosphere is the origin of these fine flavors.

For the people living the good life in their swimming pools around Los Angeles and San Francisco, it isn't so easy to imagine that the water they're chilling in once fell as snow up in the Sierra Nevada, the mountain range that forms California's eastern border. But according to the state department of water resources, 30 percent of the water Californians use actually comes from melted snow. And as much as three-quarters of their drinking water, per Jim Roche, a hydrologist at Yosemite National Park.[30]

The rainfall that comes in from the Pacific falls as snow in the higher parts of the Sierra Nevada, which reaches heights of well over 9,500 feet in places. Normally, the snow collects in the winter months, up until March/April, when the heat starts the melting process. Meltwater trickles down in the rivers or is collected in reservoirs. Over the spring and summer, the reservoirs and aqueducts from the rivers and the Sacramento–San Joaquin delta take care of water supplies, thereby sustaining agriculture and the lawns of the suburb dwellers around the big cities.

However, winter snow is not the Sierra Nevada's only water store: some glaciers also remain, although many of them have shrunk—especially after the Little Ice Age, the period of cold that ended in the mid-1800s. Some of the glaciers actually originated during this period of cold while others date further back. And there is also something people hadn't seen until recently: the so-called rock glaciers, glaciers that are partially hidden by rock and gravel but still deliver meltwater to the lowlands below.[31]

THE SERPENT IN PARADISE

BUT SOMETHING IS wrong in paradise. There's a smell of burning. Deep in the Yosemite Valley, on the western side of the Sierra Nevada, I am constantly driving through areas of charred woodland. In other places, the conifers are brown and dying after four or five years of drought. I stop and look at some destroyed trunks and note that bark beetles have run riot, the way they do when trees are unhealthy. And then, in between, I see why Yosemite is still visited by 4 million nature lovers every year: the valley that opens out between sheer mountainsides and then the magnificent giant sequoias. Mountains aren't the only things that can make you feel small.

Yosemite is the mother of all national parks. It started when Abraham Lincoln took time out from the American Civil War to sign the Yosemite Grant Act in 1864. The act granted to California areas of the Yosemite Valley and Mariposa Grove, to be preserved and used for recreational purposes by the inhabitants of the state. This move was inspired above all by the giant sequoia trees, which are found in small pockets, the largest group in Mariposa Grove. They are the world's most massive trees and among its oldest living ones, at up to 2,700 years of age. Lincoln had not seen these trees himself, but as late as the day of his death, he said he wanted to go to California to see them. Unfortunately, he never had the chance.[32]

Later, the legendary conservationist John Muir, whose achievements include starting the Sierra Club (one of the world's first and most prominent conservation groups) persuaded President Theodore Roosevelt to make a much larger area around the Yosemite Valley into a national park, thereby setting a precedent for later national parks in both the United States and other countries. If this had not happened, there would scarcely have been so many giant sequoia trees left today. However, Yosemite's national park status didn't stop the Californian authorities from building dams and large reservoirs here. The Hetch Hetchy Reservoir is one of the most important water sources for San Francisco and environs. But as became obvious a while after I was there, the construction of these dams and reservoirs did not factor in the possibility that the winter precipitation that usually falls as snow might suddenly fall as rain.

Another threat to the area is the ever-increasing stream of tourists, which now totals around 4 million visitors a year. Most of them come to the spectacular Yosemite Valley, which is surrounded by mighty granite formations like El Capitan and Half Dome. Few mountains have been photographed more often than these. In the valley itself, there isn't so much in the way of accommodations, but many of the tourists sleep out in tents, despite the presence of black bears and the occasional mountain lion. Here and there you will see signs asking you to keep food and leftovers in airtight containers so that the bears won't catch a scent of the food. I think about how different it is in Norway, where the mere rumor of a bear is enough to make people stay indoors and clamor for hunting licenses.

I see—and smell—smoke and think that a new forest fire must be burning here. But it turns out to be the forest rangers themselves who are going around lighting fires and have full control over them. They use controlled fires as a protective measure. Whereas they used to try to extinguish every tiniest fire, they have now introduced a new system. The old system caused the forest to grow twice as dense as when the woodland was left to its own devices. Under natural circumstances, fires are always starting as a result of lightning strikes, and this has served as a natural control system, renewing the forest and leaving small open patches scattered about. As a result, really destructive fires seldom happened. It was also much better for the ecosystem and the animals that lived there, as the clearings provided space for grass and other vegetation the animals needed. And in winter, snow might settle there.

This "new" practice (once widely used by Native Americans), the return of which was controversial at the outset, is largely attributable to scientists like Roger Bales. I met him in Merced, a little university town in the middle of the Central Valley, which runs north to south through California. Bales works here to find out what forest rangers, farmers, and other people can do to deal with the drought that has hit California in recent years. Here, people are no longer debating whether the climate crisis is human-caused. They're right in the middle of it and are trying to find solutions.

The increase in forest fires is *one* side of the crisis. "Forest fires are correlated with the climate," Bales tells me, "but droughts aren't unusual. What's new this time is that it is both dry *and* hot." And the fact that people have been managing the forest in the wrong way, by letting it grow so dense, has made matters worse. The more trees there are, the more water they absorb from the soil. If the forest gets too dense, the soil dries out and so eventually do the trees themselves; they then wither and are easy prey for bark beetles. And when the forest dies, it begins to release climate gases rather than absorbing them. A vicious circle is set in motion. That is why the efforts to do something with the forest here in Yosemite and elsewhere in the Sierra Nevada are so important. They make the forest better able to fulfill several important functions related to both environment and climate: to remain one of the major users of carbon dioxide but also to help Californians survive the drought.

The way to deal with the drought is to preserve and improve the water stores. And we aren't just talking about the artificial dams that they also have here in Yosemite and the Sierra Nevada, but the *natural* water stores, such as the forests and soil. Not to mention the snow that supplies Californians with meltwater in the dry months. When the forest is thinned out—something nature used to do itself through fires—more snow lies on the ground and less water evaporates.

Bales previously conducted research on Greenland glaciers, but for many years now he has been studying the hydrology (water turnover) in California, focusing in particular on what happens when the climate changes. What is new here, he says, is that the snow cover in the Sierra Nevada has diminished dramatically, reducing the amount of water in the watercourses that supply the state. This affects both forest and countryside, especially now that it coincides with higher temperatures. He says that the forest on the western side of the Sierra Nevada has been an important buffer for water, retaining a great deal of moisture. So it is a big problem that the forest is now struggling and that parts of it are dying. What is needed is forest that is healthy, but not too dense.

The drought doesn't just affect the forest. Agriculture is also suffering, and several areas of cultivated land have become unusable. At the same time, the inhabitants of this populous state must put up with tighter water restrictions. Gone are the days when people left their sprinklers running permanently; many places now have schedules that dictate when watering is allowed.

One of the causes of the water shortage is the rising population, which has grown from several thousand in the mid-1800s to almost 40 million Californians today. But at the same time, the climate has changed, irrespective of whether human activity or natural fluctuations are to blame. True, it hasn't changed that much: it has got a *bit* drier, a *bit* warmer. But this change has been enough to upset an important balance: between water as liquid and water in frozen form; between precipitation as rain and precipitation as snow.

Just how the precipitation falls is important. In a region with a Mediterranean climate, such as California, precipitation tends to arrive in concentrated periods, while other parts of the year are typically dry. And as we know, when all the water arrives at once, a lot of it will go to waste (for human purposes), even in places where dams have been built to conserve it. This was precisely what happened in January–February 2017: the precipitation fell as rain and there was too much of it for the dams to cope with. But if the precipitation falls as snow, it will act as a buffer: not all the water runs off at once. And the snow generally melts at a time of the year when people need a bit more water. Precipitation in the form of snow is therefore money in the bank: it's there until you need it.

Down in the Central Valley, the groundwater rather than forest is the most important buffer in times of drought. But agriculture here has traditionally used a lot of water, especially for what has become California's most important export: almonds. California is the world's largest almond exporter by far, and almond trees yield good income. But they demand an incredible amount of water all year round. Every tiny almond takes up about a gallon of water, and almond trees account for 10 percent of California's entire water consumption.[33] But it is profitable for the farmers, because almonds are in demand, especially among vegans who drink almond milk—perhaps imagining this helps the environment. A UCLA study shows that it takes more than 1,500 gallons of water to produce 1 quart of almond milk.[34]

The authorities are trying to gain control over water usage by introducing a new law, the Sustainable Groundwater Management Act. The aim is to get farmers to return as much as possible of the water they consume. But this is a difficult issue, involving major economic interests. California is one of the world's most productive agricultural regions and specializes in extremely water-intensive crops, such as fruits and nuts. Perhaps people will have to switch to other plants, but it isn't so easy to abandon the most profitable crops.

Despite the huge drought problems of recent years, Bales still believes it will be possible to get through this crisis. He says it is far from a given that the drought will last: such episodes of drought have also occurred before, and have ended—so the situation may reverse. The new aspect here, however, is the high temperatures, which exacerbate the crisis. People had been hoping for improvements when the El Niño weather pattern returned. And there was even a little more precipitation in winter 2015/16, following the preceding years of catastrophe. But not nearly enough. And although there was heavy precipitation in January–February 2017, it came mostly as rain and much of it ran off.

I had only just left California when that rain came. And it rained the way it often does there: cats and dogs. It rained so much that the dams couldn't hold up and the water broke through. This happened, among other places, at Lake Oroville, whose water levels are usually highest in June and not February, resulting in the evacuation of 188,000 people. And the San Joaquin River threatened to break its banks. There were rockslides and mudslides, trees fell across roads, and several deaths were reported. Even in Southern California, where it "never rains," the storms and floods claimed lives. And the cause was what Bales had spoken about: almost all this precipitation came as rain and not the snow that had been usual in the Sierra Nevada. This shift, caused by an apparently small temperature rise, was what led to *both* the flood and the drought. As Bales said: "It doesn't take much warming to change snowy weather to rainy weather. And a warmer climate will bring winter storms that dump rain rather than snow."

When the precipitation came as snow, as it did before, it didn't run off straight away, but lay waiting until spring, when it melted little by little. The snow served as a buffer. Now that it all comes as rain—because of a *slightly* higher temperature— it is all too much for the dams and water systems. They weren't designed to deal with all the water in one go. As a result, the rain that now came could not relieve the drought in the dry months. For that to happen, larger and more robust dams would be needed. People experienced this in summer 2017 when new forest fires ravaged several parts of the state. Even the heavy winter rain hadn't managed to bring the drought to an end. November 2018 brought even worse fires.

The drought problems in California show that changes in the cryosphere can have major implications even in regions with optimal, almost paradise-like climate conditions, where snow and ice never cause problems and are barely visible except to ski tourists who have opportunities to travel high into the mountains. And this is happening even in one of the world's richest and most high-tech regions. The problem of drought also shows that the climate crisis is no longer a threat that lies far in the future. It is a reality here and now, manifested as a water crisis. Such crises will also arise in other regions around the planet because there are many places where people rely on meltwater for part of the year—not least in areas around mountain ranges such as the Alps, the Andes, the Caucasus, the Karakoram, and the Himalayas. The meltwater from the last of these alone keeps hundreds of millions of people alive in the dry months of the year. And while it is possible that people will be able to deal with the crisis in California by building larger dams and better systems for conserving the water, because the finances and technology to do this are available, such solutions are not as realistic in other, poorer parts of the world where the geology is often more unstable.

A far worse problem than this, I think as I fly from San Francisco on my way northeast, is sourcing water for the enormous areas of land that spread out on the other side of the mountains: the prairies east of the Rocky Mountains, the Sierra Nevada's big brother. We are talking about the land where vast herds of bison used to wander over endless grasslands, all the way from Texas in the south to Alberta and Saskatchewan in the north. We are talking about the world's granary, which doesn't just keep the growing North American population alive but is also a vital provider of food for poor people around the world. Without grain from the prairies, food prices would be sky high and millions of people would starve.[35]

What has helped sustain this highly productive operation—other than the generous prairie soil, which is a gift from the ice age—is meltwater. Not much of this land is irrigated, and that means that water, at the right times and in the right amounts, is crucial. But changes in climate, as in California, are changing the balance of precipitation. Over the last six decades, surface air temperatures have increased. The largest warming is in winter and spring, with the result that a lower fraction of the precipitation is snow, and what snow there is melts earlier.[36] Furthermore, the situation isn't likely to get better. The latest report by the UN's Intergovernmental Panel on Climate Change (IPCC) says that it is "very likely" that there will be further decreases in northern high-latitude spring snow cover, alongside further warming that will reduce the fraction of precipitation falling as snow.[37] According to Barrie Bonsal of Environment and Climate Change Canada, it may well be that as time goes on, with more warming and associated possible increases in droughts, more farmers will need to rely on irrigation from rivers and reservoirs. And what feeds these rivers and reservoirs?

It is, to a significant degree, the flow of water from sources in the Rocky Mountains. This flow includes one of the world's largest watercourses—the Mississippi/Missouri—but also other large rivers such as the Pecos, Rio Grande, and Saskatchewan. All of these, as well as the Mackenzie and Athabasca, obtain the vast majority of their water from the Rocky Mountains, from the melting snow, the glaciers, and the lakes left by melted glaciers. Tens of millions of people across North America depend on meltwater from the Rockies for uses including drinking water, industrial needs, recreation, and sustained flows for aquatic ecosystems.[38] As in other places, people here have simply taken the water for granted. They haven't grasped that it is a resource that can, in fact, diminish because they haven't quite realized where the water comes from.

As we saw in California, if all the water comes at once, as rain, and runs straight down into the rivers, neither nature nor the current infrastructure can make use of the water. What should be a resource becomes a problem, causing floods and damage, especially in the form of erosion. If the precipitation comes as snow, however, it may remain where it falls, waiting until spring or summer and melting into water just when nature and people need it most. Or there is woodland and other vegetation that can take up the water, enabling it to circulate in the "green water cycle."[39] The water is absorbed into the plants, where some of it becomes part of the growth (as we know, the majority of plants' mass is water), while some of it evaporates and creates a separate microclimate, whereby moisture in the air condenses and transforms into new rain—as in the tropical rainforests, where the vegetation creates its own watering system. In both cases, water circulation is delayed and spread out over the year.

From the air, you get a quite different picture of the landscape than on the ground. Only when you fly over the Greenland ice sheet do you realize just how enormous the cryosphere is: white as far as the eye can see, in all directions. And when you fly over, and along, the Rocky Mountains, you see how vast this mountain range is. You can also see how water has shaped the landscape, from the glaciers on the top to the fractal, tree-like ramifications of creek beds and valleys created by meltwater from glaciers and snow. Up here in the mountains, the water follows yet another cycle: the cycle of frost. Some of the precipitation settles as snow on the mountain glaciers or on the frozen soil—the permafrost—and becomes part of the "ice box." Some of this box is invisible: we cannot see permafrost, and until now we have overlooked it in practical terms, too; we haven't understood how significant it is for water supply down in the valleys, or the complex ways its thaw changes connections to groundwater and the resulting streamflow. But regardless, the various types of meltwater feed the small creeks that join up to become rivers running down the mountainsides and valleys, where their water gives life to vegetation—from lichen and moss highest up, to subalpine meadows, and, eventually, forest and prairie.

The Rocky Mountains, which began to rise 80 million years ago, are in many ways the very backbone of North America. This enormous mountain range stretches more than 3,000 miles and is home to many of the most prototypical North American animals—like grizzly bears, mountain goats, bighorn sheep, and bald eagles. It's also the land of adventurers, where newcomers could grow rich from furs or gold in the pioneering days, but could also be destroyed if they weren't tough enough.

The mountain chain splits North America in two, especially in terms of climate. On the Californian coast, and on southern Vancouver Island, the climate is Mediterranean, with hot summers and mild, slightly damp winters. Elsewhere on the coast, an oceanic climate prevails. East of the Rocky Mountains, meanwhile (although there is also a drier, more desert-like area between the coastal mountains and the Rockies), a typical inland climate applies: hot summers and cold winters, with relatively little precipitation and large temperature shifts, generally accompanied by strong winds. The differences in climate and geography are reflected in the agriculture: conditions in the west are optimal for fruit and vegetables; on the prairies, it's mostly grain—wheat in the far west, corn farther east.

The Rockies stop the warm, humid Pacific Ocean air— the Earth's rotation means the winds usually come from the west—which condenses into clouds and precipitation on the mountainsides. Given the elevation here, where many peaks are over 13,000 feet, much of the precipitation falls as snow in winter, with several consequences: first, it has preserved many glaciers that were left behind after the last ice age, and even increased the glacial area during the period we call the Little Ice Age (from around 1350 to 1850). Second, the precipitation that falls as snow remains throughout the spring, thereby acting as a water tower that distributes water over time and keeps the rivers running longer than they otherwise would. A third consequence is that the snow has given the Rocky Mountains unique flora and fauna, adapted to life in the snow—for parts of the year at least.

For European immigrants, the Rockies were, first and foremost, a barrier where the opportunities for agriculture ended, as well as a strenuous obstacle en route to the coast, with its gold and favorable climate. What people thought less about was that the Rocky Mountains, and the water they supplied, were an important part of why it was possible to build up productive societies in this area. And that these societies existed at the mercy of the cryosphere, whether in the form of the water towers formed by the snow and the glaciers in the mountains, or the local snowpacks feeding the fields. It is difficult to calculate exactly, but some estimates of the streamflow contributions of the mountain snowpack alone exceed 75 percent in the states close to the Rocky Mountains, the Sierra Nevada, and the Cascades.[40]

The onset of warming in recent years has caused the mountain snow cover to dwindle and last a shorter time. The snow line is constantly moving farther up toward the peaks, and the snow is both arriving later in the autumn and melting earlier in spring. Consequently, although there has been no decline in precipitation, there is less water left over the course of spring and summer.

I am flying over Montana, the country's wilderness state par excellence. This is the core area of the American Rockies for both mountains and glaciers. Here, the total mass of snow has declined by 20 percent since 1980.[41] At the same time, the temperature has risen 2.3 degrees Fahrenheit (1.3 degrees Celsius)—nearly twice as much as in the country as a whole, because warming becomes more severe the higher up and farther north you go. But the picture is the same throughout the Rocky Mountains: the snow now melts between ten and thirty days earlier than before. The warming and the change in the pattern of precipitation are also reflected in the state of the glaciers. Many of them have vanished entirely since the Little Ice Age. In Montana's Glacier National Park, the name will soon be all that's left of the glaciers: two-thirds of the 150 glaciers that were there in 1850 and lent their name to the park have now gone. And those that remain are in the process of vanishing, according to glaciologist Dan Fagre, who has been in charge of monitoring the national park's glaciers for several years.[42]

The same applies to Canada, north of Montana. Between 1975 and 1998 the glacial area diminished by 22 percent in the North Saskatchewan catchment area, and by 36 percent in the South Saskatchewan. Both catchment areas originate in the Canadian part of the Rocky Mountains. Since 1985, twenty-nine glaciers have vanished in Alberta's Banff National Park. It is calculated that 25.4 percent of the glacial area in this province disappeared between 1985 and 2005. And readings show that the glaciers are now shrinking faster than before: their mass balance (the relationship between growth and melting) is becoming steadily more negative.[43]

In a way, mass balance is a glacier's balance sheet: the "income" is the snow that falls in winter; the "expenditure" is how much melts in summer. If the expenditure exceeds the income, the balance sheet turns negative, and the "capital"—the glacier—shrinks. And this is the way it has gone, on the whole, since the Little Ice Age, but with accelerated shrinkage in the past few decades. The most important factor is the rise in temperature, which has caused more of the precipitation to fall as rain and led to more melting in the summer. Here, developments in the Rocky Mountains have gone the wrong way, as in most other parts of the globe: since 1980, the global negative mass balance has tripled.[44] And in the Rocky Mountains, this has had clear results: hundreds of glaciers are gone. According to David Hik, an ecology professor at Simon Fraser University, "probably 80 per cent of the mountain glaciers in Alberta and B.C. will disappear in the next 50 years."[45]

So far the impact on water supply has not been especially noticeable. Of course melting means more water in the short term, but it is also capital that is being drained away. And later in this century, when the glaciers disappear, this will also have a noticeable effect on the discharge in the major watercourses. What will the impact then be on the many different users of the water? The glaciers' contribution may not be so great over the whole year, but it becomes important in the summer: a 2012 American study calculated that melted glacier water accounted for between 23 and 54 percent of the discharge in many local watercourses.[46] And in Canada, for example, although the Rocky Mountains occupy only 17 percent of the Saskatchewan River basin, they provide 43 percent of the annual flow in the lowlands and 80 percent of the flow in summer.[47]

If the water shortage has not yet become noticeable, the changes are already visible in another way: when the amount of snow diminishes and the temperature rises, the mountain ecology also changes. As noted, the Rocky Mountains are known for their rich and unique fauna. Some animals here cannot be found anywhere else, like the pika, a tiny relative of the rabbit. It is adapted to life high up in the mountains, where it hides in scree and under the snow. If it gets warmer than 77 degrees Fahrenheit, the pikas die out, and because the temperatures have risen in recent decades, their habitat is becoming increasingly restricted, higher up in the peaks.[48] The problem is that they have nowhere else to go from there.

A couple of the animals that prey on the pika (and other animals), wolverines and lynx, are also adapted to life in the snow: the wolverines mainly live off frozen carcasses that they dig up. With less snow, their chances of survival deteriorate, and they will probably vanish in the course of this century. The same fate is likely to befall the mountain caribou, the snowshoe hare, and the marmot, because the only place they can move is higher up—until there is no longer anywhere left for them to move.

Because the retreat of the glaciers and the changes in animal life are visible here, people's awareness of the cryosphere's decline is also growing. People are beginning to understand that popular tourist areas such as the Sierra Nevada, Yellowstone, Banff National Park, and the Columbia Icefield will change drastically over the course of this century—unless the warming stops. Even Montana's Glacier National Park, which took its name from the glaciers, may be free of ice by the middle of the century, perhaps earlier.

THE ALPS: WHEN THE CRYOSPHERE BECOMES DANGEROUS

BERNER OBERLAND, February 2018: I had been expecting snow, masses of snow. It was, after all, the worst winter for snow that the Alps had seen for ages: according to the meteorologists, it had only been this bad one other time in the past thirty years. Almost every day there were reports of villages cut off from the outside world, and tourists having to be picked up by helicopter. Ten thousand tourists were stranded in the Italian Alps. The exclusive holiday town of Zermatt by the Matterhorn was cut off. What's more, there were constant reports of avalanches claiming fresh lives. In France alone avalanches had already caused twenty deaths this season. Many of them had come in places where experienced mountain guides had never known avalanches to happen before. Even the annual meeting of the wealthy in Davos was at risk for a while until the weather eased off a bit.

So what had become of the snow? After all, I had come to Switzerland, the alpine country par excellence. This was the place where even Norwegian skiers had traveled to train when the snow failed to arrive in Norway. And it was the actual birthplace of ski tourism: although it is true that we Norwegians—or rather the Sami—were the first to do cross-country skiing, it was here in Switzerland that downhill skiing became a popular leisure activity and provided the basis for an industry worth billions.

In Norway, we traveled by ski through necessity, but the person who started ski tourism may have been the son of an English hotelier in Mürren, a car-free village in the vicinity of the better-known alpine resort of Wengen. It is claimed, in a tourism magazine I read in a Swiss hotel, that Arnold Lunn was the man who invented skiing as a sport—using skis for fun—in 1910.[49] The aim was to offer English tourists an entertaining pastime in the winter season. Lunn was helped by a local shoemaker, Fritz von Allmen, who developed a new type of shoe, which, combined with suitable ski bindings, made it possible to stand steady and zigzag down over the hills. The shoes and the bindings were known by the name of Kandahar. I, who grew up with Kandahar bindings on my skis, had always lived in the belief that this name came from Afghanistan. But the only link with the snowless town that shares its name is that a former English officer who once served there and had been granted permission to call himself the Baron of Kandahar allowed the shoemaker, von Allmen, to use this exotic name. It caught on thanks to the fact that most people knew little about snow conditions in Afghanistan's second-largest city. This is the reason why ski boots and other winter shoes are still produced under the Kandahar name—12,000 pairs a year.

But when I was there, even deep in alpine country, in Berner Oberland with the mighty Jungfrau massif nearby, there was no skiing and absolutely no snow to be seen. I traveled by bus up to the winter sports resort of Beatenberg along with a group of hopeful tourists. But no snow there either. Although it is true I could see some tourists carrying toboggans—something only children play with in Norway—I couldn't see any places where they might be used.

Still, there must have been snow somewhere, farther up in the heights, I thought, because otherwise the streets here in Interlaken wouldn't be full of Asian tourists dressed in skiing gear, mobile cameras at the ready. The Alps is the place where millions of people come into contact with the cryosphere in the form of snow or glaciers, and this has been the case for over a hundred years, ever since the English launched ski tourism. In other parts of the world, the cryosphere is still something that is remote and that few people come into direct contact with. Even most Nepalese people have never touched snow, although they can see it up in the mountains. But here in the Alps it is extremely accessible: you fly in to places with railway lines that take you high up into the mountains; for example, the line up to Jungfraujoch, Europe's highest railway station, which lies 11,332 feet above sea level and offers fantastic views of attractions such as the Aletsch, the biggest glacier in the Alps.

But it isn't just the tourists who have dealings with the cryosphere here: in the hills and valleys leading up to the mountains, farmers have cleared forest and undergrowth to create pastures for goats, sheep, and especially cows, for hundreds of years. In the mountainous areas, such as Berner Oberland, people practice semi-nomadic agriculture, which has much in common with a similar tradition in Norway: mountain dairy farming. In the summer when the snow in the hills has melted and the mountain grass is green, the farmers take their livestock into the mountains, where they graze for around a hundred days of the year. Some family members remain at home, where they cultivate hay that will be used as fodder in the winter. In this way, they can maintain large herds which they use to produce milk and dairy products. Swiss cheeses, such as Gruyère and Emmentaler, are famous all over the world. The same applies to the Swiss-cheese-based dishes, fondue and raclette. Powdered milk is another product that is sold throughout the world, especially through Nestlé, the Swiss company that has become the world's largest food company. The foundations for this success were laid in pastures watered by meltwater from snow and glaciers.

But in the Alps, as in other mountain ranges around the globe, the cryosphere is in movement. The glaciers are rapidly retreating, which often causes instability: when parts of the glaciers crack, this can create dams, resulting in flash floods or landslides. The snow cover is also changing: glacier scientist Samuel Nussbaumer, who grew up in Bern, told me that he played in snow in the streets there as a child. At that time, the snow line was at the same elevation as Bern, around 1,600 feet. But now, it has climbed to almost 3,300 feet. So that's why there was no snow to be seen in Bern, or in Interlaken, which is only slightly higher up.

There *is* snow in the Alps, though, and masses of it at that; you just have to go higher up to find it. And it also depends on where you are in the Alps. On the northern side, the Alps are relatively dry and so there is little snow. On the southern side, however, masses might fall, but levels are difficult to predict now that the weather is shifting between extremes. The many avalanches are triggered not just by heavy snowfall but also by sudden warm spells.

However, people are used to avalanches here. What is new, at least in our times, is that the mountain itself, the bedrock, is not as stable as it used to be. The permafrost has begun to shift, unleashing landslides and rockslides. In August 2017, eight people—most of them tourists—were swept away by a landslide in Val Bondasca, in the southeastern corner of Switzerland. Around 140 million cubic feet of rock and soil came hurtling down into the village of Bondo, near the Italian border. Hundreds of people had to be evacuated, some by helicopter, but the eight could not be saved.

And this was not a unique event: at more or less the same time, there was a similar slide in another valley in the same area, Val Bregaglia. Here, several tourists were reported missing but later turned up safe and sound. In this case, large amounts of rock and gravel were released from the Piz Cengalo peak. The slide was so enormous that it registered three on the Richter scale used to measure earthquakes. And twelve farm buildings were destroyed. But Val Bondasca had earlier experienced a similar slide, in 2012. After this a warning system was set up, which apparently prevented a greater disaster in 2017.

The reason such slides now occur more frequently is that the permafrost, the frozen layer uppermost in the bedrock and the soil, is now periodically thawing. The alternation between frozen and thawed states creates cracks, which fill with meltwater, and when this water freezes again, the cracks widen. The result is a constantly unstable base.

In other words, the Alps are experiencing several types of destabilization—simultaneously. Overall, there is less snow because the snow line is moving upward, but the snowfalls are becoming more intense, and rapid temperature shifts are triggering more avalanches. The glaciers are melting and cracking in many places, which can unleash floods and slides. And the fact that the permafrost is thawing creates rock- and landslides.

What makes the situation especially difficult is that these effects sometimes seem to work together and reinforce one another, explains Martin Hoelzle, leader of a group of cryosphere scientists I meet in Fribourg, a little university town on the border between German- and French-speaking Switzerland, half an hour from Bern. Here, several scientists are monitoring the glaciers and permafrost in the Alps, a job that is not merely of academic interest here in Switzerland: both the tourism industry and farmers depend on proper monitoring of the cryosphere.

The glaciers in Switzerland (and this is where most of the glaciers in the Alps are located) have been systematically monitored since 1880. At that time scientists began to take annual readings of the glaciers' length.[50] Later, the monitoring was extended to encompass glacier speed (active glaciers are like rivers in slow motion), as well as temperature, area, volume, and, not least, mass balance.

Two main aspects are decisive for mass balance: winter precipitation and summer temperature. In years when the new ice added exceeds the amount melted, the glacier will grow—and vice versa. This has varied through history: in the Little Ice Age there was a surplus for long periods, which meant that the glaciers grew—sometimes so much that villages had to be abandoned.

But since 1980, the movement has mainly been going in the other direction: glaciers have shrunk, sometimes swiftly. This is largely because summer temperatures have risen: in the years since 1980, there has only been one summer when the average temperature has been normal; in all other years, the summer has been warmer than normal, with some extreme examples, as in 2003 and 2015. When it comes to precipitation, there have been considerable variations between the north and south sides of the Alps, with much more snowfall on the southern side, although the rise in temperature has meant that glaciers have still shrunk on both sides.

Some glaciers have experienced large retreats in recent years. The largest glacier in the Alps, the Aletsch, which lies due south of the Jungfrau massif, became 104.7 feet shorter in 2013/14 and 175.5 feet shorter in 2014/15. The Rhône Glacier, also on the south side, and the most important water source for the river bearing the same name (which rises south of the Jungfrau massif, passes through Lake Geneva, and then flows through France to the Mediterranean), also retreated significantly in those years. But the greatest effects were on several smaller glaciers, such as the Kehlen, Schwarz, and Unterer Grindelwald.[51] The last of these is near populous areas, by the famous tourist town of Grindelwald, and is monitored minutely because any collapse here could cause large-scale damage and hazardous situations in this important tourist area.

The melting of the glaciers will also create risks and problems for many local communities, as well as for major industries such as tourism and agriculture. Switzerland (like other countries with alpine regions) has, for example, built up large infrastructure projects, including railways and cable cars, in the vicinity of glaciers, and even very close to them. Have people considered what might happen to them when the glaciers begin to break up, which is a highly relevant issue today?

The scientists in Fribourg were also worried about the situations that can arise when glacial melting or unstable permafrost creates dams, which eventually burst, causing damaging floods and slides. The impact of such complex events is extremely difficult to forecast, but these are precisely the kinds of events we can expect more of in a future where the climate fluctuates more dramatically than before.

The glaciers are visible and relatively easy to monitor, but the permafrost is a different matter. And there's a lot of it in the Alps, for example in rock glaciers—glaciers that look like the remains of a rockfall at first glance, but which contain large quantities of frozen water and move exactly like a glacier (just a bit slower). That is also probably why people weren't especially concerned with them until pretty recently. Whereas systematic monitoring of the glaciers dates back to 1880, scientists didn't begin to monitor the permafrost until 1999. So in this case, the changes are less certain, although all indications are that permafrost is also on the retreat.[52] The most important factor is the rise in atmospheric temperature, but the ever more extreme shifts in temperature and precipitation are an even more crucial factor for permafrost than for the glaciers. When meltwater seeps into the cracks and spaces and then freezes, this breaks up the bedrock even more, setting rocks and gravel in motion. (As we know, water expands when it freezes—something you may have experienced when you've tried to cool a can of beer quickly by putting it in the freezer. And then forgotten it.)

So the shrinking and increasingly unstable cryosphere is a factor of the greatest uncertainty for people in the Alps. But what about those who live farther down and depend on the water from glacial rivers, such as the Rhône? Will people experience more drought when, as is likely, the water supply from these gradually diminishes? This is, in fact, something that is already noticeable in certain areas of Switzerland, especially the southwestern canton of Valais, according to Hoelzle. But for now, it seems as if the weather gods (i.e., the climate systems) are on humanity's side and that there is still plenty of water in the areas around the Alps—sometimes a bit too much.

Even so, the events of recent years—the rockslides, avalanches, and cut-off villages—have shown that people in the Alps are at the mercy of the cryosphere. It has enabled humans to live up here in the mountains, but not much in the way of climate change has to happen for it to become a dangerous neighbor.

But there are areas of the globe that are at least as dependent on the local cryosphere—the glaciers and snow cover—without knowing that ice and snow exist there.

— 11 —

THE RIVER GODDESS AND HER SISTERS

WHILE I WAS living on the western coast of Norway, where snow and frost were rare visitors, I wasn't especially concerned about the cryosphere. It wasn't something that affected me. Paradoxically enough, it was only when I spent some time sweltering in South Asia, far from the Arctic, that I "rediscovered" the Kingdom of Frost—although not directly, because there is neither snow nor ice on the Indo-Gangetic Plain that covers much of northern India and Bangladesh. Very few people there had ever seen anything like that. Even in Nepal, snow was just something you might glimpse on the peaks in the Himalayas, and you'd have to be a mountain climber to get up there.

It was when the Indo-Gangetic Plain was at its hottest and driest that I realized the significance of the cryosphere. In the hot, dry months before the monsoon sweeps in from the Indian Ocean—which usually happens in June—people here are utterly reliant on the water that still runs in the rivers, thanks to the glaciers and snow in the Himalayas. If they only had the rainwater, they could not survive. Nothing would grow in the dry fields, the wells would run dry, and people and animals would die of thirst. Even now, they were living at the very edge of the possible. What little water they could get hold of for household purposes they—that is to say the women or girls—often had to walk hours to fetch. Only a tiny minority could afford to build deep enough wells, and, what's more, the groundwater was polluted with arsenic in many places.

But although water was so important, people were not aware of where it came from. They thought all of it came from the sky as rain, which fell into the hair of the river goddess, Ganga (i.e., the forest), and then ran down toward the Bay of Bengal. And for some months of the year that was true, of course. The monsoon brings a great deal of rain: Cherrapunji in the Indian state of Meghalaya (due north of Bangladesh) is known as the wettest place on the planet. Even so, in the hot spring months, meltwater from the snow and ice is what stops the rivers from running dry. The rainwater runs rapidly out into the Bay of Bengal as there is nothing to retain it. And it generally takes a great deal of soil—and many houses—along with it.

But on the Indo-Gangetic Plain, few people had heard of meltwater; even the term was unknown. People there had never seen snow or ice and had barely even heard of them. Or rather, *some* had heard of snow and became very curious when they found out I came from a country where we spent parts of the year living in it. "What does it feel like?" "Is it cold?" "Can you eat it?" But these eager questioners had no idea that snow and ice had anything to do with them and that the cryosphere was, in fact, what kept them alive for several months of the year.

My first stay in South Asia was in northwest Bangladesh, where my partner was doing fieldwork for a dissertation about marriage customs in a particular tribal group, the Oraon. They were a hospitable people with an egalitarian culture, originally forest dwellers who now lived as farmers and agricultural workers. But they had preserved some hunting traditions, even though there was barely any forest left to hunt in. They still used bows and arrows, mostly to scare off intruders who might try to encroach on their territory. They lived in a kind of truce with the Muslim majority, who viewed them as heathens and barbarians. The "worst" thing was that they kept pigs and ate pork, as well as brewing a dangerously strong rice beer, which they often found occasion to consume in copious quantities.

The tribal population, the Muslims, and the few Hindus who remained in Bangladesh after the partition of India all made a living from agriculture: wheat in the winter, rice in the rainy season, and other than that, a few vegetables and domestic animals, such as buffalo, cows, goats, pigs (only the tribal people), chickens, and ducks. Incidentally, this applied only to those who could afford to keep such animals; for most people, it was rice, rice, and nothing but rice. Those who did keep animals used absolutely every part of them, just like *flyttsamer* with reindeer: when they slaughtered a pig for a village festival, everything, including the entrails and trotters, was carefully distributed because it could be put to good use one way or another. Even the pigs' ears were carefully fried to collect the fat they gave off. There was no rubbish: if any empty bottles or tin cans showed up, they were also used for some purpose.

Most people probably associate Bangladesh with masses of people living in hunger and poverty, especially those who remember TV images from the difficult 1970s, when the country was hit by the Liberation War and cyclones. The aid concert arranged by George Harrison, in particular, did much to paint a picture of Bangladesh as the very epitome of misery. And it is true that the country has suffered many catastrophes. But Bangladesh is actually a fertile country; it just has a few too many people: 164 million in a region the size of southern Norway. They benefit greatly from the monsoon that comes in from the Indian Ocean every early summer, which brings cooling and refreshing rain, and which can sometimes be pretty intense. The same goes for the cyclones in late summer. But no less importantly, much of the country consists of the delta formed by two of Asia's mightiest rivers, the Ganges and the Brahmaputra, as well as a third river, the Meghna, which is also big but not as far traveled—coming mainly from the mountain ranges in the border areas toward the northeastern "appendix" of India. So you might think there would be plenty of water, and indeed there is, for much of the year. Every year floods cause great problems: they destroy roads, sweep houses out into the rivers, and leave people homeless. Lives are also lost.

But in the spring months, before the monsoon generally arrives—sometimes it does not appear at all—large expanses of Bangladesh suffer drought. This also applies to the region where we were staying. There are many reasons for this, the most important being that the discharge of the rivers varies tremendously and dries up when the rainwater has run off after several dry months. There are also few water storage systems, partly because so much of the land consists of loose delta soil, which is easily displaced when there is flooding, making major infrastructure projects difficult. The rivers generally seek out new courses, making riverboat traffic a nightmare, and the boats often run aground on newly formed sandbanks. Building dikes is also a demanding and risky business. The same goes for bridges, which makes car travel hazardous. Even where there are roads, a bridge may suddenly turn out to be missing.

The problem with water shortages is greatly exacerbated by the fact that the two most important rivers, the Ganges and the Brahmaputra, run through India—the Ganges from the northwest, the Brahmaputra from the northeast—before reaching Bangladesh. India has its own water shortages and therefore takes its fill of the water, even though an agreement on sharing the Ganges (which divides and then runs out into both West Bengal and Bangladesh) is supposed to ensure that the Bangladeshi people receive a share at least. But India is the one in control here.

The importance of water—how it determines the day's activities, from trips to the communal well with water pitchers in the early morning, to cooling dips in the tiny village pond in the late afternoon, and what can be grown on the dry patches of land—is reflected in religion too. For most people along the Ganges water system (that is, Hindus), the river goddess Ganga is the most important divinity. She is the goddess they worship above all others because she is the one they depend on. And all the major events in life take place on the riverbanks and out in the river in itself. What's more, the ashes and partially burnt remains from cremations are thrown out into the holy river. The goal of all Hindus is to end their lives as ashes in the Ganges or one of her tributaries. In holy cities such as Varanasi you can see people bathing in the river—a rite of cleansing—as mortal remains float past.

I became so fascinated with this that I planned and made an entire television series about the Ganges water system.[53] In the series I covered the way that the Ganges and her tributaries colored the existence of all the millions who lived near her, from the Himalayas to the Bay of Bengal. It was a pretty time-consuming project because travel is not easy in this region and I didn't have a large budget. We traveled partly by jeep, sometimes by boat, and, in the mountains, mostly by foot with plenty of help from porters. But this allowed me to experience a little of what the people in this region have to struggle with, and the project certainly made its mark on me—not just because I lost a lot of weight, either. It was a bit difficult to come back to Norway and try to engage with the trivial issues that concern people here after seeing how millions of the world's poorest people—the Indo-Gangetic Plain has the planet's highest concentration of people in poverty—strive to keep their lives together. The intrigues of soap operas like *Dynasty* (which was very popular in Norway back then) seemed even more meaningless.

One scene that is still etched in my memory is something I witnessed in a dried-out riverbed in Bihar, India's poorest state. I saw a gaunt woman in ragged clothes walking along searching, blade by blade, for grass she could cut with a small pair of rusty scissors: food for the emaciated goat that accompanied her. The goat appeared to be her sole possession. And almost no grass grew there. A bit farther on sat a little girl, presumably the woman's daughter, who might not get any food other than the few drops of milk the goat could yield.

What I didn't manage to record in this TV series, although I gradually began to grasp the scope of the issue, was how much of the discharge from these rivers, the very lifeblood of the people, depended on meltwater in the dry months—in other words, just how important the cryosphere was for people down here, in the hottest and most populous region of the planet. This applied to 300 or 400 million people in northern India and Bangladesh alone, and I knew the same applied to most people in Pakistan (with its 180 million inhabitants), whose own lifeblood, the Indus, is extremely dependent on meltwater. It also applies to China because the biggest rivers there come from the highland regions of Tibet, where there are glaciers, snow, and permafrost. Indeed, the glaciers and snow in the Himalayas/Karakoram/Tibet serve as water towers for at least *one-fifth* of the world's population, more than a billion people. And regardless of whether the rivers run out east in China, south in Pakistan or the Bay of Bengal, or perhaps even in Southeast Asia (Mekong and Salween), their source lies in the same region.

Although the cryosphere is remote for most people down on the Indo-Gangetic Plain, they still have some concept of it.

In their religious accounts, the origin of the life-giving water is up in the mountains, in the white. That is why the mountains are holy, the abode of the gods. And the route up to them follows the rivers, especially the Ganges.

THE CENTER OF THE UNIVERSE

"NO, TIGERS ARE not available here, sir," the driver says, with that characteristic, slightly stilted word choice typical of Indian English. No, there are apparently no tigers in the forest we are driving through on our way up into the Indian Himalayas. We are aiming for the place where the Hindus' holy river, the Ganges, rises. My photographer and I have rented a car with a driver from my usual New Delhi base, the slightly shabby York Hotel. The driver is well acquainted with the Ganges valley, not just where tigers can be found. We stop in the holy cities on the way up the mountains: first Haridwar, where the river runs so swiftly that the pilgrims have to hold on to a rope when they walk out into the ice-cold water to cleanse body and soul. And then Rishikesh, where the Beatles once went to study with the Maharishi at his ashram on the other side of the river. Both the Beatles and the Maharishi are long gone (Paul McCartney once said the only song they wrote during their stay was "Ob-La-Di, Ob-La-Da"), but the ashram still stands. We wander up there, but they do not want a visit from TV people. They have enough publicity and more than enough nutters trying to get in. What the city most resembles is a religious bazaar, where the sale of Hindu souvenirs is more conspicuous than contemplative asceticism. The two evidently coexist perfectly happily together.

The holiness of the Ganges is emphasized by the fact that both Haridwar and Rishikesh are "pure" towns: you cannot get hold of either meat or alcohol. However, in a little area between these pilgrimage sites, there is a small free zone, where the drunkards rave by the roadside, if they haven't already lain down to sleep there. The holy life probably isn't for everybody.

However, the most zealous pilgrims continue upward along the holy river. If both time and physique permit, their goal is the source of the Ganges, which, according to Hindu tradition, is the Gangotri Glacier up in the Himalayas. Here, too, there are other possibilities, other sources you can walk to. In Mustang in Nepal, I once visited such a source, reduced to a little water tap from which one could eke out a few scarce drops of water. Perhaps I visited at the wrong time of year. But behind this pilgrimage business lies an ecological truth: the rivers of the Ganges water system are absolutely vital to the population and must therefore be treated with respect. And their origins lie up in the ice.

Our journey does not take us to Gangotri. In those days that would have required weeks of travel by foot (although there is now a road), and as we shall see it isn't even certain that this really *is* the source of the Ganges. Our goal is the place where the Ganges officially begins. This is by Devprayag, where the rivers Alaknanda and Bhagirathi meet. Only from this point is the river called the Ganges, so of course Devprayag is also a pilgrimage site.

There aren't many people there when we arrive, just a few hardy sadhus (holy men) who walk out into the ice-cold water at the precise spot where the rivers join. These kinds of meeting points (*prayag*) are often holy: the place where the Ganges and the Yamuna meet in Allahabad is the site of the world's largest pilgrim gathering, Kumbh Mela, which is held every twelfth year. But we see that Devprayag is also used to visitors; a little way uphill from the river stands a concrete building with a slightly terrifying name: "complex toilet." Just a problem of word order, one hopes.

While religious people claim that the Bhagirathi is the actual origin of the Ganges, since it begins beside what they consider to be the source of the Ganges, hydrologists have something quite different to say: for them the Alaknanda is far more important. First, it is much longer, and second, it has a greater discharge, so most of the water in the upper part of the Ganges derives from it. The Alaknanda also comes from glaciers: the Satopanth and Bhagirath Kharak Glaciers in Uttarakhand. But the religious focus on the Bhagirathi has the advantage that it has made it easier for engineers to build hydropower stations along the Alaknanda.

Confusion about the actual source was exacerbated in 2015 when India's minister of water resources, Uma Bharti, asked the national hydrology institute to investigate whether the actual source of the Ganges lay somewhere else entirely—by the holy Mount Kailash, which has a connection to Ganga and Shiva in Hindu beliefs.[54] The hydrologists were obliged to follow up on the minister's request, and promised to investigate it by tracking radioactive isotopes in the river water.

It is fairly doubtful whether the Indian hydrologists will really be able to trace the water of the Ganges back to Kailash, which is in western Tibet. But if they did, the history of the rivers from the Roof of the World would be complete—almost a religious/hydrological "Theory of Everything." After all, it is in or close to Kailash and the nearby Lake Manasarovar (also spelled Mansarovar)—one of the world's highest lakes—that the other great rivers of Southeast Asia have their origin.

The world's fourth-biggest river, the Brahmaputra, runs east. It follows almost the whole of the Himalayas, initially flowing on the northern side of the world's mightiest mountain chain, only to take an abrupt turn to the southwest, into northwest India, and eventually ending up in Bangladesh, in the world's largest river delta. Toward the west runs the Sutlej, first through India, before entering Pakistan and eventually joining the Indus. Toward the north flows the Indus, which turns southwest along the way, becoming the very lifeblood of Pakistan. It also gave rise to one of the world's oldest civilizations in the Indus Valley. Toward the south flows the Karnali, which cuts through the Himalayas and then through Nepal before descending to the Indo-Gangetic Plain and joining the Ganges. So regardless of what conclusions the Indian hydrologists reach, we can already say that much of the water in the Ganges comes from Kailash—or, as it is known in Tibetan: Gang Rinpoche, "a precious jewel with snow."

Both Hindus and Buddhists view the mountains on the Roof of the World as holy. In some Hindu legends, Himalaya (also called Himavat) is a god, the father of Ganga the river goddess and of Shiva's wife, Parvati (Shiva is also associated with the mountains). Himalaya lives on a mountaintop with his queen, Mena, in a palace decorated with gold, waited on by maidens and magical beings. His name is formed from the words *hima* and *alaya*, meaning "home of the snow" in Sanskrit, the ancient Indian language still used by Hindu priests.

The body and abode of the god Himalaya is a kind of reservoir of frozen water, and the divine source of many holy rivers, including the Ganges and the Indus.

In Buddhism, which is a religion without gods in principle (though not in popular practice), there is a greater focus on the holy mountain Sumeru (which the Hindus call Meru). In the real world, Sumeru manifests itself as Kailash and, as well as being the source of several important rivers, it is considered the center of the universe. Making a pilgrimage to it is therefore like visiting the very hub of what the Buddhists call the "wheel of life."

Other, more local pilgrimage sites have their own stories: in the Indian Himalayas, which lie a little closer to the Hindus than the remote Kailash, the mountain called Nanda Devi ("the goddess Nanda") serves something of the same function. Many pilgrims also flock to the mystical source of the Ganges, Gomukh ("cow mouth") on the Gangotri Glacier, and the temples of Shiva and Vishnu in Kedarnath and Badrinath respectively. Traveling up here by foot is a life goal for many sadhus.

In Nepal, the mountain goddess is Annapurna, a manifestation of Parvati. She lives in Annapurna I, one of the mountains in the mighty Annapurna range, the full circuit of which many pilgrims aim to walk once in their lifetime. *Annapurna* means "she who is filled with food"—apparently a reference to the mountain as the source of one of the Ganges's major tributaries, Kali Gandaki ("the black Ganges"). Annapurna is also called "the queen of Benares" (i.e., Varanasi), the holiest of the holy cities along the Ganges.

Unlike the Norse mythology about Niflheim, these stories prevail to this day and people live by them. They worship the gods and goddesses of the mountains and make pilgrimages to the holy mountains and sources (the holy source of Kali Gandaki is in Muktinath, Nepal). And in Kailash, this remote mountain in Tibet, the religions converge. Despite the lack of roads, persevering pilgrims from *four* different faiths find their way here: Hindus, Jains, Buddhists, and followers of the original Tibetan religion, Bon. These last are distinguished from the others in that they travel around the mountain in a counterclockwise direction. But going around the mountain is what all the pilgrims are supposed to do, in any event, assuming that they still have enough energy left after their long journey here.

I have never been one to view expressions of old religious tradition as eternal truths, but in this case, I must almost accept that the pilgrims are right: what they are seeking really is the source of life. The four named rivers that come from here (and since Karnali is a fairly important tributary, we can include the Ganges too) really are the alpha and omega for hundreds of millions of people in Asia. The Indus water system, with its irrigation canals, is the main artery of Pakistan, and in the dry season (i.e., for the greater part of the year) most of its discharge is meltwater from glaciers and snow.

The Brahmaputra plays an equivalent role across large stretches of Tibet, then later in China, northwest India, and Bangladesh. Even these last regions need river water after the monsoon has worked its devastation, flooding rice fields and roads. As I have said, even the river country of Bangladesh experiences drought. And the Ganges makes life possible for the hundreds of millions who live on the Indo-Gangetic Plain, which stretches across the whole of northern India. Here, in a place that is home to the largest concentration of people living in poverty, the inhabitants live at the mercy of the river goddess, at the very margins of the possible.

The mountains themselves are not the source of the water, however; it comes from the glaciers on and around them. For the region is so dry that almost no rain falls here, so the water that is the origin of all four great rivers is old water, collected over thousands of years. The west winds do bring a bit more at certain times of year.

Now these sources, the mountain glaciers, are in danger. There wasn't much talk of this in the 1990s when I was making television programs about the Ganges and her sisters. Only later did we see signs that the ice on the Roof of the World was in the process of shrinking. But these signs were somewhat contradictory and the consequences, if it was true, were too enormous to take in. If hundreds of millions of people were left without water every year, where would they go? All previous refugee crises would pale by comparison. At that time, in any case, people could only speculate about the outlook: no firm data were available. That would come later.

Moreover, the climate issue had not yet come to the fore in those days. That only happened after scientists began to conduct serious studies into what was happening in the core regions of the cryosphere, at the poles. As a science journalist in the 1990s, I kept track of this, particularly because I had contact with the ever-growing climate science circles in Bergen, western Norway. The starting point for this science was ocean research, particularly in the Arctic.

TOWARD NIFLHEIM

Shrouded in fog lay the mythic land of Nivlheim. ... There in the darkness and cold stood Helheim, where the death-goddess held her sway; there lay Nåstrand, the shore of corpses. Thither, where no living being could draw breath, thither troop after troop made its way. To what end?[55]

N JUNE 1893, a little ship set out on an expedition that most considered to be sheer lunacy. They were traveling in waters nobody had sailed before, and worse still, the crew were going to let the vessel freeze into the ice just off the coast of Siberia and simply hope that the ice and ocean currents would carry it over the Arctic Ocean, across the North Pole to Greenland. The best indication they had that this plan might work was that some timber thought to have originated in Siberia had washed ashore in Greenland. Foreign experts thought the project was both technically untenable and highly risky, but the Norwegian parliament and the Norwegian king both provided funds for the hazardous undertaking, which would hardly happen these days. The fact that Norwegian official-dom granted funding to the project says a great deal about the high status the expedition leader, Fridtjof Nansen, enjoyed in those days.

This was not the first hazardous expedition Nansen had led. Some years before, he had crossed the Greenland ice sheet on skis with a team of five that included two Sami men, whom Nansen considered to have special expertise in coping with Arctic conditions. They reached the eastern coast of Green-land in a whaling vessel, the *Jason*, on July 17, 1888, and were dropped off there in two small rowing boats. The journey to the coast nearly ended in catastrophe: drift ice almost crushed them on several occasions and they drifted far south, scarcely managing to struggle ashore. After almost a month (!) the team were finally ready to start their skiing expedition. The first part of the journey must have been incredibly tiring, as they had brought heavy equipment, which they first had to haul up inclines of up to 6,500 feet, each man pulling a sledge weighing over 200 pounds. Once they were up on the ice sheet, it became a bit easier, and this is where they truly came to experience the scale of this icy wasteland. In his book about the trip, Nansen wrote: "Imagine six tiny mosquitos marching across a tremendously big bed-sheet. ... There was snow and nothing but snow wherever you turned your gaze."[56] When at last they reached the west coast, the year's last ship to Europe had left and they had to overwinter there.

The journey across Greenland won Nansen the status and support he needed for his even bolder, but also much more important, expedition across the Arctic Ocean. He was actually a scientist, after all; in fact his most important contribution before he devoted himself to the Arctic had been in the area of neurology. The aim of the expedition across the Arctic Ocean was to find out how the ocean currents and the ice behaved up there. This enabled Nansen to lay the foundations for modern polar research—especially ocean research—and several research centers have been named after him. But he was undoubtedly also an adventurer. And the expedition on the little ship *Fram* was a proper adventure: a journey into the absolute unknown, with many dangers and surprises lying in wait.

Nansen's two perilous expeditions launched research into two major elements of the cryosphere, whose origin and history are widely divergent. The skiing expedition across Greenland involved crossing the largest *glacial* mass in the Arctic, Greenland's inland ice. There is no need to cross it on skis to experience the scale of it: it offers an imposing spectacle even when you travel over it by plane. It began as grains of snow many millions of years ago and has survived several ice ages and interglacial periods. In many places, it is still nearly 2 miles thick. Because it has lain there so long, it is an invaluable archive of climate history which Danish scientists, among others, have put to good use, conducting ice-core drilling thousands of feet deep.

The Greenland ice sheet contains so much ice that sea level would rise 23.3 feet if it were to melt entirely. The only ice mass larger than this one is in Antarctica. Until now, the Greenland ice sheet has been seen as stable, and it was not thought that global warming would be able to melt it even over many millennia. However, the developments of recent years suggest that melting could happen considerably faster than previously assumed.

The *Fram* expedition traveled through the *sea ice*, the second major element of the cryosphere and much more transient: it is formed when seawater freezes, expanding and contracting in tandem with the seasons. Some of this ice, around one-third of it, remains throughout the summer and may have been there for many years. The total mass of ice in both summer and winter has decreased in recent years, becoming one of the clearest signs of global warming. The sea ice is also one of the least stable elements in the context of climate change, because melting sea ice can trigger a number of unclear processes: a diminished albedo effect and changes in ocean circulation. Some would say that the very key to the future of the climate lies here in the Arctic sea ice. But this was not a problem anyone contemplated in Nansen's day.

Nansen had personally instructed the renowned ship-builder Colin Archer on the construction of the *Fram*. The important, and unusual, feature of this specially designed Arctic vessel was that it must be able to withstand being frozen into the pack ice. This is sea ice that is packed together and moves in such a way that it would crush any normal ship to matchwood and literally screw it down under the ice. The noise of it alone was terrifying, as Nansen himself described:

> For when the packing begins in earnest it seems as though there could be no spot on the earth's surface left unshaken. ... In the semi-darkness you can see it piling and tossing itself up into high ridges nearer and nearer you—

floes 10, 12, 15 feet thick, broken, and flung on the top of each other as if they were feather-weights. They are quite near you now, and you jump away to save your life. But the ice splits in front of you, a black gulf opens, and water streams up. You turn in another direction, but there through the dark you can just see a new ridge of moving ice-blocks coming towards you. You try another direction, but there it is the same. All round there is thundering and roaring, as of some enormous waterfall, with explosions like cannon salvoes. Still nearer you it comes. The floe you are standing on gets smaller and smaller; water pours over it; there can be no escape except by scrambling over the rolling ice-blocks to get to the other side of the pack. But now the disturbance begins to calm down. The noise passes on, and is lost by degrees in the distance.

This is what goes on away there in the north month after month and year after year.[57]

However, the *Fram* (whose name, appropriately, means "forward") managed to withstand the pack ice so effectively that it was used for subsequent polar expeditions, including Roald Amundsen's voyage to the South Pole. So the repeated encounters with the planet's most dangerous predator, the polar bear, posed a greater risk—and there were plenty of close shaves. Like when Nansen and his traveling companion Hjalmar Johansen left the *Fram* and set off across the ice on foot to try to reach the North Pole (an attempt they had to abandon). On one occasion, a polar bear caught Johansen off guard—an episode he later described in his account of the *Fram* expedition:

It gave me a wallop on the right cheek with its vast forepaw that made my skull rattle, but fortunately I did not faint. I fell over onto my back and there I lay between the bear's legs. "Get the gun," I said to Nansen, behind me. I could see the butt of my loaded gun in the kayak beside me, and clawed with my fingers, trying to get hold of it. I saw the bear's maw gaping right above my head, its fearful teeth glistening. As soon as I fell, I got hold of its throat and I held that grip with the power of desperation. The bear was a bit taken aback by that: this wasn't a seal but some other strange creature it wasn't used to, and to that hiatus I most certainly owe my life. I waited for the shot from Nansen and noticed the bear looking over toward him. It seemed to be taking a long time as I lay there, and I said to him: "Right, sir, you'll have to hurry up now, otherwise it'll be too late."[58]

The fact that Johansen managed to keep up the formalities, calling Nansen "sir" even in this critical situation, is confirmed in Nansen's own account of the story. Nansen explains that it took such a long time because the weapon was in the kayak and he didn't dare shoot from there, for fear of hitting Johansen.

But it is hardly surprising that the polar bear fancied a spot of human meat. Their principal food source is seals, which occasionally surface in the open channels between the ice floes to breathe. Seals, meanwhile, live off fish and crustaceans, which abound up there in the Arctic Ocean, especially at the ice edge—the point where the ice meets the open sea. This area is particularly rich in nutrients and supports one of the world's most important stocks of fish: the cod that migrate between the Barents Sea and the Lofoten archipelago of Norway. It is therefore viewed as a highly vulnerable area when it comes to activities such as oil and mineral extraction. As a result, this is a hotly debated issue, to the extent that even the definition of "ice edge" is a tricky political topic.

After giving up on their attempt to reach the North Pole (they got as far as a latitude of 86 degrees and 4 minutes), they found their way to Franz Josef Land, where they overwintered. The next summer, they were lucky enough to meet a British expedition and traveled to Vardø in northern Norway on their ship. By then, they had spent three years in the Arctic Ocean, including three winters. The rest of the members of the expedition, who had continued the voyage with the *Fram*, returned to Skjervøy in northern Norway a week later. After this, Nansen was a national hero and an international superstar, too. And many still try to copy his exploits, especially the skiing expedition across Greenland, admittedly under much easier conditions. It is even possible to take the trip as an "adventurer," accompanied by a guide and without having to struggle with all your baggage yourself, as long as you have plenty of money.

But Nansen was the first, and his most important legacy is not that people follow in his ski tracks, but that he paved the way for Arctic research, in particular the large cryospheric elements: the Greenland ice sheet and the Arctic Ocean around the North Pole. These are connected, forming the great white cap in the north that we can see on that famous photograph taken from outer space.

— 13 —

THE WHITE CONTINENT

THE ARCTIC VESSEL *Fram* did not reach the North Pole with Nansen. Even so, the *Fram* expedition was a milestone and led to a breakthrough in our understanding of the Arctic. And many years later, the same ship would enjoy yet another triumph in an even more renowned expedition: the famous race for the South Pole between the Norwegian Roald Amundsen and his men, and the British explorer Robert Falcon Scott and his party.

There are several reasons why this is perhaps the most legendary race in history: it was a hazardous undertaking in the world's most inhospitable environment—the icy waste of the Antarctic. What's more, the region was not properly mapped, so nobody knew exactly what challenges lay in wait along the way. We should also recall that in those days there were no satellite telephones or GPS and no airplanes to take the members of the expedition home if they got stranded. So when they set out, it was with their lives at stake, and it ended in tragedy for one team—with just as much glory going to those who won: Roald Amundsen and his four collaborators.

The prelude to the race was also like something from a thriller. Everybody believed Roald Amundsen and the *Fram* were heading for the North Pole. Their official aim was, in fact, to complete what Nansen had tried to do: travel up to the Arctic Ocean and let the drift ice carry them as close as possible to the North Pole so that they could reach that goal. It was on these grounds that Amundsen had obtained Nansen's permission to use the *Fram*, and this was also the purpose for which he had obtained funding from the Norwegian parliament. It was only when they lay off Madeira—the crew must have been wondering what they were doing so far south—that Amundsen dropped the bombshell: they weren't traveling to the North but to the South Pole. However, none of the members of the expedition withdrew, and the voyage south could continue.

The change of direction can hardly be put down to a whim. Amundsen was a thorough planner and was extremely well prepared for the expedition. He had already spent one winter in the Antarctic with the *Belgica* and had also been the first to travel through the Northwest Passage, in 1906, in another famous polar vessel, the *Gjøa*. This voyage took nearly three years, and along the way, he had learned from Inuit how to survive extreme cold. Inuit had also taught him to drive dog teams, a skill that would be of decisive significance for the expedition to the South Pole.

But one other contributing factor was probably that two Americans, Frederick Cook and Robert Peary, both claimed, independently of one another, to have been to the North Pole, in 1908 and 1909, respectively. And for Amundsen, being first was important. The scientific aspect—which was part of the aim of the originally planned expedition—probably came second to that. And since the North Pole already appeared to have been "conquered," only the South Pole remained.

But Antarctica was an unmapped continent, and that made Amundsen a "proper" explorer, not just an adventurer. Antarctica is not just cold; it is also so isolated and remote from the rest of the world that nobody had even seen it before 1820, when a Russian expedition observed what later became known as the Fimbul ice shelf. But the first landing on this inhospitable continent may have been at Cape Adare in 1895 (by the Norwegian-Swedish whaling ship *Antarctic*). And the first people who spent the night there, as I've said, were two Sami men. The Kingdom of Frost in the north had come to the Kingdom of Frost in the south.

Antarctica is the giant of the cryosphere: a whole continent filled with ice. And it isn't a small continent, either: at 5.4 million square miles, it is almost twice the size of Australia and 1.3 times as large as Europe. And almost all of the continent—98 percent—is covered in a mile-thick layer of ice. There is so much ice here that it accounts for 98 percent of all the ice on Earth, and sea level would rise by over 180 feet if this ice were to melt.

The cold has kept human activity at bay, thereby preserving Antarctica as one of the world's last wildernesses. Most of the people who (willingly!) live here—between 1,000 and 5,000 depending on the season—are scientists: biologists, geologists, physicists, atmospheric scientists, astronomers, and glaciologists, not to mention climate scientists. Gradually, aided by technology such as ice-core drilling and radar observations from satellites, they have begun to form a picture of the history of this icy waste.

Antarctica is a product of the great continental drifts that have reshuffled the landmasses of the world across the ages. Once upon a time, nearly all of the continents were joined, and there have also been times when what is now Antarctica lay in much more hospitable climes, as evidenced by the diversity of its fossils. So even Antarctica, cold and untouchable as it may seem, has not always been the way it is today.

Antarctica did not begin to freeze until around 30 million years ago, when it separated from Australia and New Guinea, and an ocean current formed around it, isolating it from the rest of the world. The pace stepped up considerably when the Drake Passage opened up between Antarctica and South America 23 million years ago. At the same time, perhaps as a result of new weather patterns, the carbon dioxide content of the atmosphere fell and it became colder. Ever-larger stretches of the continent froze, and eventually, 15 million years ago, it became almost entirely covered by ice.

Antarctica is not just the whitest continent, it is also the coldest: this is where the lowest temperatures on the planet have been recorded. The Russian research station Vostok, which lies 11,444 feet above sea level on top of 2.5 miles of ice, holds the record. On July 21, 1983, a temperature of –128.6 degrees Fahrenheit was measured there, and the average annual temperature is around –70 degrees. There are no months in which the average daily temperature is higher than –22 degrees, and the highest temperature measured is 10 degrees.

Few organisms thrive in such a climate. The largest land animal in Antarctica is a flightless midge measuring not even a quarter inch in length. Certain penguin species, such as emperor penguins, do live there, but they are partly aquatic animals and get their food from the ocean. The sea around Antarctica, the Antarctic Ocean, is in fact one of the most productive on the planet. Huge quantities of krill provide the basis for abundant animal life, especially when it comes to sea mammals. This was where whaling developed into a veritable Klondike in the 1900s, with Norwegians as some of the most efficient whalers—so efficient that they nearly wiped out the entire stock, even of the world's largest animal, the blue whale.

On land, though, it is quite a different matter. Although this continent is so inhospitable, certain people have still made their way to these endless, ice-cold plateaus. And even returned there. It must have something to do with the dimensions, with the fact that there is no other place where you will ever experience so much whiteness, in all directions, as here. It may also have something to do with the fact that the enormous quantities of ice and snow give an impression of something eternal: a fixed reference point in an otherwise changing world. We have all seen how the ice in the Arctic is melting, with pictures of polar bears clinging to ever-shrinking ice floes, but things have been different in Antarctica. On this frozen continent, where only emperor penguins and the hardiest scientists overwinter, the ice sits fast at astonishing depths. And nothing has previously suggested that it would vanish. On the contrary: until recently, we received reports that told us the ice was expanding, at least in the eastern part, home to the world's largest ice cap.

When Amundsen and his party went to the Antarctic, they knew that the British explorer, Scott, was on his way and that it would therefore turn into a race. So speed was of the essence, and more efficient logistics may have been the decisive factor for the Norwegian team. Whereas Amundsen had learned how fast dog teams can travel and relied on sled dogs (118 to start with), Scott had opted to rely mainly on motor sledges and ponies. It would be a fateful choice: the diesel engines froze on the very first day and things went little better with the Shetland ponies. In the end, the men were left alone with their sledges, which were also too heavy: three times heavier than those of the Norwegians. On their way back from the pole, dejected at having seen they were beaten by Amundsen, they were plagued by even worse luck: Scott's planned meeting with supporting dog teams from the base camp failed to take place, and Scott and his party died. Amundsen opted instead for dogs on the journey to the pole, and in a pretty brutal fashion: when they had made their contribution, they were used as food. In any event, they were most useful in the climb up to the Polar Plateau. The way back was more downhill.

But Amundsen made his share of mistakes. After they arrived in January 1911 and had prepared for the trip by constructing stores containing 6,700 pounds of provisions and equipment some way across the ice, Amundsen wanted to start the final attempt a week into September. He thought the Antarctic summer was on its way, but he met with resistance on this point from one of the members of the expedition, Hjalmar

Johansen, famous for his part in the *Fram* expedition with Nansen. Johansen thought they should wait. But Amundsen chose to rely on his own judgment, and things almost ended in catastrophe when it suddenly turned cold again, close to –67 degrees Fahrenheit. They got back alive, but this disagreement made Amundsen feel that his authority was threatened, so Johansen was not allowed to come along on the expedition to the South Pole.

Only the outermost edges of the Antarctic terrain were mapped, so neither Scott nor Amundsen knew exactly what lay ahead. In many places, what looks like a stable covering of ice and snow is actually a "living" glacier, in other words moving ice. Glaciers are like slow-moving rivers: they head out toward the sea, where they calve or melt. This means that in certain places the ice has crevasses that can be mortally dangerous. Later Antarctic expeditions have experienced this, including one led by Monica Kristensen, who wanted to follow in Amundsen's "ski tracks." The expedition had to be curtailed when one of the members of the party fell into a crevasse in the glacier and died. Other expeditions have also had to turn back.

So the members of the Amundsen expedition appear to have been lucky as well as skilled. They kept clear of crevasses and loose ice blocks and found a route up a glacier that took them onto the Polar Plateau relatively quickly: an ascent of nearly 10,000 feet. Up on the plateau it is relatively flat, and in order to find out when they had reached the actual pole, they had to make calculations based on the sun's position in the sky. Thus they were able to plant the Norwegian flag there, although they would have to wait almost three months to tell the rest of the world about it, in a succinct telegram dispatched from Hobart in Tasmania: "Norway's flag planted at the South Pole. All well! Roald Amundsen."

THE VOYAGE TO the South Pole is generally read as a race between two "heroes," the winner, Amundsen, and the tragic but heroic Scott. However, it can also be read—metaphorically at least—as a lesson in how extreme conditions, and especially the cryosphere, can force adaptations that enable living organisms to cope with these conditions. In this case, an adaptation took the form of collaboration. Not just between people— Amundsen was utterly reliant on his team, which consisted of men with complementary qualities—but also interspecies collaboration between people and dogs. (Some may point out that the collaboration was fairly one-sided, given the outcome for the dogs.)

British scientist Richard Boyle has concluded that extreme cold can create new forms of collaboration. Boyle's initial area of interest was how single-celled organisms that existed early in Earth's history appear to have been forced into multicellular forms of collaboration, the beginning of complex life, during glacial periods—snowball Earth events. When the climate later became more livable, they were able to use the new advantages they had acquired in order to reproduce rapidly and outcompete their less developed peers.[59]

Boyle's teachers and colleagues, Tim Lenton and Andrew Watson, have taken this thought process further and believe that extreme cold has played a similar instigating role later in the history of life. They like to use the Antarctic, and even Amundsen and Scott, as examples: "This idea can be tested by going to one of the most extreme environments on today's Earth, Antarctica, and examining how organisms survive there. Often it is by teamwork. Consider the human pioneers that first explored this great wilderness—Amundsen, Scott, Shackleton—they didn't go alone—their only hope for survival was as part of a team, and even then individuals, and in Scott's case a whole team, perished."[60]

Lenton and Watson write that Boyle's theory has a clear prediction that can be tested: that the first individuals in new lines of species will have appeared after global glaciations. They believe there are good examples of this, like the "Cambrian explosion," which took place 542 million years ago—"just" after the last snowball Earth episode. At that time, an unusual diversity of species sprang up—both flora and fauna.

Another example Lenton and Watson highlight is the emperor penguins. These, the world's largest penguins, are also the only ones that overwinter in Antarctica on the ice and they manage it thanks to unusually close collaboration at two levels. One level is the flock, which stands densely packed together for month after month, to shelter one another and keep warm in the face of the brutal winter storms. Nobody who has seen the French film about these marvelous birds, *March of the Penguins*,[61] can have failed to be captivated by their perseverance and especially by the second level of loyal collaboration, that between the females and males. While one of the parents waddles off to the coast to catch a fish supper— much longed for after months without food—the other must stand guard over the egg, week after week, until respite comes. This is the only way they can produce new emperor penguins. The least hint of sloppiness, a few days too many out in the tempting ocean, and their partner will perish, along with their offspring. Thanks to this demonstration of faithful behavior, the emperor penguins became role models in the American Bible Belt.

Behind this extraordinary behavior lies a ruthless selection, whereby only the most steadfast, faithful, and persevering are able to reproduce and pass on their genes to new generations. And since, apparently, it was not so easy to develop new physical adaptations—in many ways the emperor penguins were probably as well adapted as they could be—the only alternative was to evolve extremely unselfish and persevering behavior. A close collaboration that required 100 percent reliance on one another was the key to survival. And it was the frozen world, the cryosphere, that brought out these qualities.

But the emperor penguins are just one example of how life has developed an endless series of remarkable adaptations, under pressure from the constant new challenges imposed on it by a changing cryosphere. Earth has been a "laboratory" for the development of what is, to the best of our knowledge, a unique phenomenon in the universe: life—in ever new and ever more complex forms. And the cryosphere has always played a decisive role.

The Cryosphere Today

The factor that determines the size of the cryosphere is which areas experience temperatures below the freezing point of water over the course of the year. This criterion applies to 35 percent of the Earth's surface, including around half of the land surface.

When we talk about the cryosphere, we generally think of the "eternal" ice, the glaciers. These cover 10.8 percent of the land surface, and the largest ones are the Greenland and Antarctic ice sheets. Of the total 6.2 million square miles of glaciers, the Antarctic accounts for 5.3 million and Greenland for 672,000. The remainder, about 200,000 square miles, is spread across the globe. An even larger fraction of the "eternal" ice belongs to the Antarctic and Greenland if you factor in volume or weight, because those ice sheets are so thick. The ice in the Antarctic accounts for 6.09 million cubic miles of the 6.86 million total, and the ice in Greenland for 696,000. As I've noted, if all the ice in Greenland melted, sea level would rise 23 feet; if the same happened in the Antarctic, there would be a rise of 184 feet.

The glaciers are not the only form of permanent frost. Permafrost also covers a large area: 8.8 million square miles, although it is smaller than the glaciers if measured by volume. It covers 15.4 percent of the Earth's land surface, mostly in Siberia and Canada (as well as islands in the Arctic Ocean), and can be over a thousand feet deep.

The variable cryosphere is an especially interesting factor for the climate and can trigger strong feedback effects. I am referring here to sea ice and snow cover, but also those parts of the permafrost that vary with the seasons. Sea ice is the one people have focused on most, especially in the Arctic. The shrinking ice around the North Pole has become a symbol of climate change. The reduction in sea ice opens up opportunities for shipping, as well as the extraction of oil and minerals, but it also creates a dangerous feedback effect: the increase in open sea in the Arctic means that the ocean will absorb more solar energy, while less will be reflected back into the atmosphere. This accelerates warming, especially in the Arctic, where the temperature is climbing more quickly than anywhere else on the planet. The most recent reports indicate that Arctic Ocean ice may be totally gone in just a few decades, for much of the year at least.

The situation in the south is a bit different, because in this case a continent, Antarctica, covers a large area around the South Pole. Although little of the sea ice here is multiyear, it forms over a larger area in winter than in the north, reaching a maximum of 5.6 million square miles in September.

Snow cover is an even more variable part of the cryosphere. The variations are greatest in the northern hemisphere. Most of the snow in the southern hemisphere is in the Antarctic, where it snows all year round, whereas large stretches of the northern hemisphere only become white in winter. The seasonal snow cover in the northern hemisphere can extend over 18 million square miles—nearly half of the land surface there.

This means that snow cover is the most important variable in the cryosphere, especially when it comes to the albedo effect (see "A Closer Look: Albedo" in chapter 3), which is a major will completely disappear. So it is possible to isolate parts of nature and make them into a laboratory.

The Earth can also be seen as a laboratory, where natural experiments are continuously taking place, unplanned by anybody and often extending over long periods of time, but which we can nonetheless study, especially after the event. As our planet follows its orbit, solar radiation varies and even the tiniest shifts in this respect can trigger major changes on Earth, because it is not a uniform globe but is composed of different and changing systems: a hot and turbulent core and a surface that is slowly but surely being reshaped; a place where ancient continents move about and sometimes collide, and where ocean currents and wind patterns are in a state of constant change; not least, it is a place where large amounts of water freeze, melt, and evaporate in an eternal but varying round dance.

In its early phase, Earth was pretty much bombarded by comets, asteroids, and other "rocks from the heavens." As analysis of the 67P/Churyumov-Gerasimenko comet has shown, the ingredients of life, at least, came to Earth from outer space. Once the building blocks—vital minerals and amino acids—were in place, together with the water in which they could develop, the stage was set for one of the most spectacular natural experiments in the universe: evolution. This is something that may only have happened a single time in a single place in the universe; there is, at any rate, no sign that anything similar has taken place in our vicinity—and we are talking light years here. And all along the way, the frozen world, the cryosphere, has played a decisive role. All the important breakthroughs in the history of life here on Earth appear to be connected to events in the cryosphere. The Earth has been like a gigantic laboratory for the development of life, as well as ever more complex life forms, and the cold has been a catalyst.

Earth has undergone a long series of highly dramatic climate changes: from being entirely free of ice to being totally covered in ice (snowball Earth episodes), as well as various intermediate stages, with corresponding variations in humidity and storminess. It is these changes, which radically altered living conditions, that made the Earth into an enormous laboratory for developing different life forms.

As shown by the fossil studies of the Norwegian biologist Kjetil Lysne Voje, among others, life forms undergo constant change, not least through random genetic mutations. But only now and then do these variations lead to lasting alterations. Most mutations are footnotes in history because they are detrimental to the organism and as a result they are not carried forward and do not spread.[62] For changes to become lasting, something must generally occur in the environment, or somebody must move, and as a result certain mutants suddenly gain an important advantage.

Like when humans migrated north from Africa and were consequently exposed to less sunshine. Their bodies produced less vitamin D and many of them fell ill. But those who developed paler skin as a result of a mutation benefited more from the sunlight and therefore got by with less sun. As a result, those with pale skin gained an advantage and gradually came to dominate the northern population.

In that case, it was the organisms that moved, but it can also happen that the landscape alters as a result of tectonic shifts—like when the Indian tectonic plate clashed with the

Eurasian plate, resulting in the formation of the Himalayas. What's more, there have been changes in the climate that have caused dramatic alterations in living conditions, and continue to do so. The periods in the history of life in which old species have died out and new species have emerged have largely been connected to periods of dramatic climate change. And most of them have been linked to major changes in the cryosphere.

The history of our own species is very closely linked to changes in the climate and in particular to the changing relationship between the cryosphere and other geophysical systems. We know that the family of animals we belong to, the great apes, began to develop at a time when the Earth's climate took a chillier turn, when the polar regions began to freeze in earnest at the beginning of the last ice age period. A similar glaciation was the backdrop when the human genus— *Homo*—separated from its forebear *Australopithecus*. And also when our own species—*Homo sapiens*—emerged.

It may seem odd that events in the Arctic and the Antarctic could affect what happens in Africa, which is where the great apes and, eventually, humans came into being, after all. But when large amounts of water freeze in the polar regions, taking the form of both land and sea ice, much of the moisture in the atmosphere becomes tied up; given feedback effects, this also causes a general cooling in the sea. The result is a decline in precipitation, which causes droughts and changes in habitats. So, yes, when ice comes and goes in the Arctic, it sends ripple effects south to Africa.

The freezing of the polar regions, and later, also, of North America, northern Europe, Siberia, and large mountainous areas, caused the climate to change across the entire globe. In

Africa, it led to a drier climate, which drastically altered the vegetation. In particular, we know that something like this happened at the precise time when our forebears (who were sneaking around as "southern apes," *Australopithecus,* in those days) embarked on a period of rapid biological change after a relatively long period of stability, which resulted in the emergence of several new species over a short period. The cause was adaptation to a new climate and a new environment: the forest thinned out and "we" had to get used to the African savanna. The fruits of the forest were much scarcer here and there was also much less protection from predators. On the other hand, there was more of a different kind of food—meat from herbivores—and good opportunities if you could manage to find ways to catch them without being eaten yourself.

All the evolutionary steps leading to modern humans coincided with changes in the climate and cryosphere, and especially with glaciations. So in purely biological terms, we appear to be "children of the ice."

But it doesn't end there. Since we became *Homo sapiens,* around 200,000 years ago (some new findings suggest 300,000, but this is uncertain), we haven't managed to undergo many biological changes, but we have, on the other hand, changed in other ways: socially and culturally. Not least, we have developed new technologies and new types of social structures: societies. And here, too, the same applies: the great leaps happen in connection with movements in the cryosphere and climate change: the transition from hunter-gatherer to settled agricultural society; the first city-states and civilizations; modern democracy and the industrial society. All these changes coincide with events that we can read from ice-core drilling in the Greenland ice. It may sound peculiar—but there is a possible explanation for it. And it lies in the mechanism for the development of life that Darwin discovered.

As I noted in chapter 7, the emergence of living organisms—and in particular oxygen-producing cyanobacteria—altered the atmosphere of the Earth. This created the conditions that led to the first glaciation, possibly 2.9 billion years ago, and a series of ice ages half a billion years later.

During the period of relative stability that followed, the oxygen level continued to rise and life forms slowly adapted to an existence in which oxygen was a resource and not—as originally—a poison. And at some time during this period, the Proterozoic, a remarkable thing happened that would alter the history of life: a meeting between two single-celled organisms, probably an archaeon and a bacterium. Apparently, one of them tried to eat the other, but they ended up entering into a symbiotic relationship instead, thereby forming the first eukaryotic organism—that is, a cell with a nucleus that would ultimately make it possible for multicellular, and therefore more complex, organisms to form. All multicellular organisms that live on Earth today are descended from this fateful meeting. But it did not immediately lead to any major dissemination. That would take yet another snowball Earth event—an occurrence that life itself helped bring about, because the net effect of life on Earth is cooling. Without life, it would be about 81 degrees Fahrenheit warmer.

But didn't these glaciations put a stop to life? No. We know that simple, single-celled organisms and even eukaryotes can be frozen and thawed again. Some very tiny creatures, like tardigrades, can also withstand such treatment.[63] And even when Earth was completely covered in ice, there were still hot springs down in the ocean bed, where many now believe life originated. Nonetheless, these glaciations did put the organisms to the test, creating what biologists call selective pressure, which causes some to perish while only the best adapted survive. And what can increase the likelihood of survival under conditions of extreme cold? According to biologists who have studied life in cold climes: collaboration.

We can see this among the few animals that spend the winter in the Antarctic, like the emperor penguins. We also saw it among the first people who went to that continent to try to "conquer" it, such as Roald Amundsen's team: alone, they wouldn't have stood a chance in the icy waste. But the same also goes for much smaller organisms. Collaboration makes existence easier, and for certain single-celled species, this collaboration became so close that they developed into multicellular organisms. After that they could also specialize, dividing the tasks among themselves, so that some cells dealt with sense perception, others with digestion, others with defending the organism against external enemies, and so on. Of course, what I'm talking about here is not planned collaboration, but something that came about through evolution, where random coincidences that turn out to work are retained in new generations.

Such collaborative projects, which were the start of what we call animals and plants, apparently arose as emergency solutions to extreme conditions—and the global glaciations must certainly have been extreme. But then, when the Earth thawed again, the field—or sea—was open to these new creatures because so many of the others had perished. And now they could spread out. There have been many examples of this dynamic in the history of life: a difficult climatic episode, generally a glaciation, selects for new organisms, which "take over" when the climate improves and creates opportunities for expansion and diversification.

And the same also went for what may be the most spectacular example of evolution we know of: the "Cambrian explosion."[64] It happened some 540 million years ago, "just after" the slightly dull Proterozoic, which concluded with the last snowball Earth event. At that time, a diversity of multicellular organisms "suddenly" appeared in three biological kingdoms: plants, animals, and fungi. The explosion involved not just an enormous increase in biological diversity but also the emergence of life forms that were far more complex than previously: animals and plants with advanced systems for converting energy, materials, and information.

It is from the Cambrian explosion onward that evolution as Darwin described it becomes clearly visible, with its biological diversity, as well as new species that come and (for the most part) go. Now, there are no longer any total glaciations, which would probably finish off most multicellular species. Still, this doesn't mean that glaciations no longer take place at all. On the contrary, it is in later geological history, once the continents have fallen into place, that the world settles into its rhythm of ice ages and fluctuations between cold and warm periods. This rhythm of constant climate change gives evolution a series of "pushes," in which old species have to give up the ghost and new ones emerge. But now it never goes so far that the whole globe becomes covered in ice, so there are always some zones left where there is liquid water and where complex organisms can survive.

What happens when the climate changes is that the vegetation also alters, thereby transforming animals' habitats. This happened in Africa, when the climate began to grow colder around 50 million years ago. As more and more ice formed in the polar regions, the climate in Africa became drier and what had once been forest became grassland. Only then did the typical African savanna landscape appear, providing food and space for millions of herbivores, from elephants and rhinoceroses down to antelopes and smaller animals. And in their wake came all the predators: lions, hyenas, wild dogs, leopards, jackals, and cheetahs, thereby completing the food chain: predators (and parasites) eat plant eaters, which themselves eat... guess what. But the greediest of all the predators was yet to come.

At the same time, the animals that had lived in the great forests, like the apes, were exiled to ever-smaller areas, while some developed new adaptations to enable them to survive at the forest edge and in more open terrain. Some of these were our forebears, the first hominins. They constituted a new branch on the tree of evolution, one that separated from the chimpanzees—both geographically and genetically—around 7 million years ago. While the chimpanzees remained in the forests, the hominins headed out onto the savanna. Most of them perished, but *one* species would eventually conquer the world.

— 15 —

CHILDREN OF
THE ICE

LISABETH VRBA IS a Yale professor and one of the world's foremost paleontologists (fossil scientists). She is best known for having developed the theory of exaptation in collaboration with Stephen Jay Gould. According to this theory, many traits and qualities useful to species today may originally have served quite different purposes. For example, birds' feathers were not originally particularly suited to flight but acted more as temperature regulators. Somewhat less known is another theory that Vrba developed alone and which did not initially attract so much attention, although it has gained fresh relevance today because of climate change. It is the theory of turnover pulses, large changes in the diversity of species, which seem to occur simultaneously in a large range of species, and generally coincide with climate or other environmental changes.

Vrba's theory took root when she was a young scientist in South Africa studying antelope fossils. She found that the fossils were particularly abundant in limestone caves of the Transvaal, and it was possible to identify and sort them by age. When she did so, she made a remarkable discovery. A striking change had occurred among the fossils that were around 2.5 million years old: "Anatomical features such as teeth indicated that antelopes living before 2.5 million years ago had occupied moist woodlands. Shortly afterward, however, these forest antelopes disappeared and were replaced by many new species that graze only in dry, open savannas."[65]

It wasn't just the antelopes that had undergone such comprehensive change, and scientists later found similar changes in other parts of Africa too: in Ethiopia, Kenya, and Tanzania. It was clear that something dramatic had happened to the climate at that time to make southern and eastern Africa notably drier and transform the forests into savannas. The climate change in Africa was part of a larger, global shift. It was precisely at this time, around 2.5 million years ago, that the northern hemisphere experienced the culmination of an extensive glaciation and the beginning of what we call the great ice ages. Antarctica was already covered in ice and now the same thing was happening in the northern hemisphere. The glaciers grew and spread: first over Greenland, then across northern Europe and Russia, and finally across North America, where the largest ice cover would remain. But the effects were not limited to the northern regions. The whole global climate was changed. It became colder and drier because the glaciation extracted so much of the moisture from the air and greatly reduced sea level.

In turn, there was a dramatic change in the vegetation. It was now that the typical savanna landscape so familiar to us from parts of Africa came to dominate large areas previously covered in forest. The animal life changed along with the vegetation: instead of animals adapted to life in the forest, species developed that were better suited to the new grasslands and, like the oryx, were best adapted to dry conditions. The changes came gradually, through anatomical and behavioral adaptations, but after a time both flora and fauna were transformed. This didn't just apply to the antelopes: other steppe animals, such as horses, now appeared in Africa.

At the same time, several new species of hominin came into existence—the group of which we are the sole remaining species. First came several variants of our early ancestors, *Australopithecus*. One of these eventually became the first species in the *Homo* line, of which there were later several versions and whose history must constantly be revised. But the first ones apparently included *Homo habilis* and *H. rudolfensis*. *H. erectus* was more like us and was also the first in the *Homo* line to migrate from Africa and spread to the Middle East, Asia, and Europe. Our nearest relatives, the Neanderthals and Denisovans, may have descended from *H. erectus* or from *H. heidelbergensis*, which is also possibly our own immediate ancestor.

This happened at a time when the world was undergoing climate change that led to ever colder and drier conditions in Africa. All these new species succumbed pretty quickly, with the exception of one species from which we originated. Vrba thinks these kinds of dramatic, comprehensive evolutionary changes have happened several times in connection with climate change, and she calls them turnover pulses. The climate is what instigates these evolutionary changes because the environment and living conditions are so drastically altered that new adaptations become necessary. A new kind of food demands new teeth, a new digestive system, and a new pattern of behavior. This is why herbivores (plant eaters) are different from frugivores (fruit eaters).

What often happens in such situations is something that Harvard biologist Ernst Mayr described as early as the 1940s: large climate changes or other environmental shifts can cause some groups of animals or plants to become isolated from other members of their species, for example if the landscape is split up into "islands." Then they will no longer exchange genes and will therefore generally evolve in different directions. If this goes on for long enough, they become so genetically dissimilar that they can no longer reproduce with one another. They have become different species.

Changes in climate and environment are therefore factors that drive evolution, leading to the development of new characteristics and, eventually, new species. Under stable conditions, however, not much happens: then the populations will be in so-called stasis, an evolutionary balance in which mutations continue to occur but are generally rapidly eliminated.

So in the period analyzed by Vrba, new antelopes were not the only ones entering the field; the first species in our own line, *Homo*, were also emerging. And there was a particular cause for this: a dramatic growth in the cryosphere that resulted in a colder, drier climate. The start of the great ice ages, from around 2.5 million years ago, was also the start of our own species' line. Without this glaciation, there is no guarantee that humans would have appeared. We are children of the ice.

When Vrba launched her theory about turnover pulses, it met with skepticism, perhaps because it was perceived as deterministic to draw too direct a line between natural conditions and human history. A lot of people would rather not believe that nature has a direct influence on human destiny. But later research has confirmed the main points of her theory. The British geographer Mark Maslin and his colleagues are among those who have studied how the natural conditions in East Africa changed at the point when the hominins emerged, from around 7 million to 1 million years before our time.

This was a period when the glaciations in the northern hemisphere were intensifying, causing drought in Africa. The record shows an increase in sand and dust from the Sahara and Arabia, among other things. But there were also considerable climate fluctuations in this period and these coincided with geological changes in the Great Rift Valley, which runs through the whole of East Africa, from north to south. One result of these changes was that lakes formed in the Great Rift Valley, lakes which came and went. Maslin and his team think this may have been among the factors that determined how the earliest human species came into existence and how they spread across the landscape. Their research draws a more detailed picture of the highly variable landscape in early human history than was possible when Vrba proposed her theory, but they nonetheless support her principal points. The

British scientists conclude that the changing climate during this period (from the appearance of the first hominins to the time when *Homo erectus* emigrated from Africa) played a significant role in the evolution and spread of the early human species. The increasing glaciation in the Arctic had a decisive effect on the climate in eastern and southern Africa, as well as on human evolution.[66]

But why did these changes in climate and vegetation lead to us humans, in particular, being the species that survived? Rick Potts of the Smithsonian Institution has an interesting theory: what happened when the ice ages set in, from around 2.5 million years ago, was that the climate changes came more quickly and more frequently than before. This favored species capable of adapting, which required them to both have larger brains that enabled them to produce and use tools, and travel long distances. It was precisely these factors that would distinguish us from the other hominins. We rapidly developed a bigger brain, which made it easier for us to adapt to changed circumstances, find new survival strategies—not least by collaborating in groups—and use simple technology such as stone implements and fire. We also developed hands with fine motor skills that were able to handle and use this technology, as well as arms and shoulders that could be used to throw stones and eventually spears. And the fact that we walked upright meant that we could move over larger distances, even out of Africa and into the rest of the world. In other words, the very fact that the climate was now changing so often was the reason why a hominin species with modern human traits was selected: a species that could think, throw, and walk or run a long distance.[67] Both javelin thrower Andreas Thorkildsen and chess grand master Magnus Carlsen have their distant forefathers and foremothers in Africa to thank for their skills.

It was not just our human genus—the *Homo* line—that originated in a period when the cryosphere dominated the Earth's climate (around 2.5 million years ago). Our species *Homo sapiens* also came into existence during an unusually cold and dry era. This was in an ice age period when there was merciless cold in the north and ice cover over much of Europe, Siberia, and North America. In Africa, there wasn't so much snow—only on the mountaintops, where a little still remains; but on the other hand, it was dry there, the way it is during ice age periods, when a lot of water is tied up in ice.

It was here, on the ever-drier African savanna, that the first *Homo sapiens*—modern humans—appeared. It is impossible to say exactly when the very first one emerged because we haven't yet found any traces of them. New findings suggest that some of them already existed 300,000 years ago, but this is a little uncertain. Nor do we know how many of them there were, although genetic analyses suggest that there were only around 10,000 at one point. This was a demanding time for our forebears too, and, as appears to be the rule, it is in just such times that new variants come to the fore. The others fail to adapt to new conditions, and so either their species line dies out or a new variant manages to survive by creating a new niche for itself.

And what seems to be a tendency in the evolution of our species is an ever-improving capacity to make use of new types of technology. This started with *Homo habilis*, who appeared at the same time as stone axes, and continued with *H. erectus*, who had more advanced tools and the ability to control fire. With *H. sapiens* came more types of technology that left our species even better placed to tackle the frequently brutal climate swings that accompanied the ice age period. These new technologies, such as clothes, domestic implements, and better weapons, probably helped *sapiens* fare better than their relatives, the Neanderthals and other descendants of an earlier *Homo* species, when they eventually went out into the wider world (the Middle East, Asia, Australia, and Europe); perhaps these technologies also helped them outcompete or eradicate those hominins, who had been living in Europe and Asia for several hundred thousand years. It seems we also had more highly developed language, meaning that we could communicate and collaborate better.

Homo sapiens left Africa during an ice age, when northern Europe and parts of Russia were covered in ice. But south of the ice cover, on what was then tundra, conditions were favorable for large plant-eating animals such as reindeer and mammoths. The to-and-fro movement of the ice actually created fertile soil that was good for growth of grass and lichen, which these animals could graze on, the way they still do (minus the mammoths!) in northern areas of Scandinavia, Russia, and North America. For the human species, which mastered big-game hunting and could survive in the cold, these were good, if demanding, times.

We can still see traces of this big-game culture in the wealth of impressive cave art to be found from Spain in the west to the Urals in the east, art which has survived for more than 20,000 years. The motifs were typically drawn from the life of the big-game hunters, depicting many scenes of animals and hunting. Nobody knows exactly how these cave paintings were used or what kinds of ceremonies surrounded them, but renowned archaeologist Steven Mithen thinks the cave paintings played a central role in a cultural/religious practice whereby knowledge about animals and hunting was shared among human beings.[68]

In order for people to survive as big-game hunters, diverse and detailed knowledge was vital: which animals it was worth hunting, when and where they migrated, which places it was favorable to attack them, and how this should be done (traps, collective strategies, weapons, etc.). It was necessary to learn how to make effective weapons, all the way down to which raw materials should be used. Knowledge and skills were also needed when it came to surviving in a brutal climate, in which many predators saw you as a potential meal and in which other people, of your own species and others, perceived you as a dangerous competitor or an enemy. Living as a hunter in the ice age was tough, but for those who managed it, it could be a good life. Even today, after all, plenty of people pay a fortune for the opportunity to enjoy a taste of such an existence—going big-game hunting or setting out into the wilderness to live a "primitive" life.

Yet nobody pays for a chance to live like the big-game hunters' descendants, the first farmers. You don't see ads for "adventure holidays" as guest workers in Vietnamese paddy fields (farm tourism is about tourism, not work). The farmers' lives weren't just a great deal more boring, they were also less healthy, as we will see in the next chapter.

Even so, it was the transition to fixed settlements and eventually agriculture (the cultivation and harvesting of agricultural plants, and animal husbandry) that would lead to what we call civilization. And that dramatic change in human life—this time more cultural than genetic in nature—also had a connection with, and was probably triggered by, changes in the cryosphere and the resulting climate fluctuations.

— 16 —

OUT OF EDEN

S FAR BACK as the old ones could remember, life had been good there in Abu Hureyra. The weather was warm and pleasant, interspersed with lovely cooling showers of rain. The forest was filled with nutritious nuts: pistachios they could eat just as they were, growing in long rows of trees. Acorns needed a bit of work to make them edible but were good and filling, and there were masses of them in the forest. At the forest edge grew grass with seeds that people would later call "grain": wild variants of wheat and rye. In the forest, there were also animals they could hunt. In summertime especially there was a surplus of meat when large flocks of desert gazelles passed through, offering easy pickings.

The combination of these favorable conditions made it possible for many people to live in the settlement we now know as

Abu Hureyra, which was excavated by the British archaeologist Andrew Moore prior to the construction of a hydroelectric dam in northern Syria. The people of Abu Hureyra had it good, perhaps better than anybody before them ever had. They had it so good that they didn't need to develop any new technology or new survival strategies. This really was a Garden of Eden. And the same was probably true for many other people in the then so favored part of the world that we now call the Levant.[69]

But then suddenly, about 12,800 years ago, the climate changed abruptly. The life-giving rain showers disappeared; it became drier and colder and the forest was no longer as generous. The forest also shrank. This created problems for the relatively large settlement of Abu Hureyra, because now there was no longer enough food to go around. All they had left was the wild grain varieties, the origin of what would become wheat and rye. But it was difficult to get enough nourishment out of them, so people began to look for new solutions. They had seen that grain could grow when it was buried or scattered on moist earth. Some people began to do this on purpose, using seeds from the varieties of grain that gave the highest yield. They also chose plants that didn't shed their seeds as soon as they ripened and which were therefore easiest to harvest. Sheer need forced the inhabitants of Abu Hureyra to develop this further. We say that "necessity is the mother of invention." We can also say that necessity taught these people a new way of producing food.

After a few generations, these new practices had altered the grain varieties at the genetic level. Without knowing it, the people of Abu Hureyra had carried out the first plant breeding experiments and had now become farmers. They were no longer hunters and gatherers who simply helped themselves to nature's bounty. They had altered nature, breeding plants that they cultivated on land suited to the purpose. They had no idea they were embarking on what may be the greatest revolution in the history of humanity: the agricultural revolution.

And this didn't just happen in Abu Hureyra: archaeologists have also found traces of the same development in other places in the Levant and the Middle East. In Turkey's Karacadağ Mountains, geneticist Manfred Heun, a professor at the Norwegian University of Life Sciences in Ås, has studied how the population bred new variants of einkorn, a wild relative of wheat, at around the same time as the people of Abu Hureyra were starting to cultivate rye and wheat. Heun and his colleagues have analyzed the genes of both cultivated and wild varieties of einkorn and discovered what was apparently the forerunner of modern wheat.

According to Heun, einkorn is extremely rich in nutrients and also lends itself to beer making. And this may be more significant than previously thought, says Heun: he thinks beer played a role in the establishment of the world's oldest temple complex, Göbekli Tepe, located close to Karacadağ. He believes beer was brewed to mobilize the workforce needed to build this complex, which was not directly linked to any settlement. Earthenware jars were found at Göbekli Tepe that were too big for grain storage and which were probably used for brewing. Heun himself brews einkorn beer.[70]

People in the northern Levant, also known as the Fertile Crescent, were fortunate in that variants of grain, pulses, and other plants grew here that could be altered through breeding to become useful crops. And, as they would discover a bit later, there were also species of animal here that could be domesticated, such as goats and sheep. But they had coexisted with these plant and animal species for millennia without domesticating them. Why did this agricultural revolution happen at just this time?

The explanation lies in the dramatic events taking place in the Arctic, involving the cryosphere, which altered the climate all the way down in the Middle East. It was as if somebody switched off the Gulf Stream. Quite suddenly, this huge "heating cable" stopped sending warm seawater—and therefore also warm weather—to the North Atlantic and northern Europe. The immediate cause was that so much cold fresh water settled on top of the sea in the North Atlantic that the mechanism which keeps the Gulf Stream going—thermohaline circulation—came to a halt, totally or partially. The warm salt water that comes from the south normally sinks and returns southward far below the surface, "sucking" new water from the Caribbean. But when a "lid" of fresh water builds up at the place where this process happens, the mechanism stops working. The water on the surface doesn't sink, causing the "suction" to vanish and the stream to halt.

People are still debating what caused this sudden flood of fresh water. The prevailing theory is that fresh water suddenly started flowing out of the Gulf of St. Lawrence in what is now Canada from a large lake, Lake Agassiz, which formed as a result of glacial melting in North America. (The lake previously ran out into the Gulf of Mexico.) This sent huge amounts of fresh water out into the North Atlantic, and this water settled like a lid on the surface of the sea, bringing thermohaline circulation to a halt. Another theory is that the polar jet stream moved north, also as a result of the changes in the ice cover, and that this led to a new weather pattern over the North Atlantic involving a lot of rain. This again resulted in a layer of fresh water on the surface of the ocean.

Regardless of the original cause, the impact on the climate was dramatic. Temperatures fell several degrees, and for more than a thousand years, until about 11,500 years ago, it was as if the ice age had returned. In northern Europe, forest was replaced with tundra. The dryas, an arctic flower that has lent its name to this period, bloomed again. But farther south, too, the climate change made itself felt. True enough, it didn't become especially cold in the Middle East, but the problem there was that it became so much drier. After several thousand years of a warm, humid climate with good growing conditions for plants and, therefore, also animals—perhaps the best people have ever had—times suddenly became much leaner. The forest retreated and people who were used to helping themselves to its fruits, such as nuts, berries, and animals, had to try to find other things to eat. For many, it ended in starvation, but others who were a bit luckier and perhaps a bit smarter led humanity into a new way of life: agriculture.

The Younger Dryas event, more than a thousand years of drought, created the first farmers in the Fertile Crescent. When the climate improved again, people continued to use the new agricultural methods and the new species that had been bred. They didn't return to hunter-gatherer society even though it was healthier and less labor intensive in many respects. Once people have adopted a new technology, they don't "undiscover" it and go back to life as it was before. Besides, agriculture made it possible to feed a lot more people.

But it took a dramatic climate change caused by a major event in the cryosphere to force us to become farmers. Perhaps it is the memory of this event that is reflected in the Bible story of God punishing humans by expelling them from the Garden of Eden and forcing them to work the land, which was evidently perceived as a more exhausting existence: "In the sweat of thy face shalt thou eat bread."

However, this was just the first step on the road to civilization. These first farmers were still not social citizens in the modern sense. They lived in family groups, were largely self-sufficient, and didn't have a great deal of contact with one another. And over the next 3,000 to 4,000 years, this was a good way of life, on the whole. The climate was optimal: the cold, dry northwesterly winds had stopped. Instead, the westerly winds brought humidity and warmth to the Middle East, where the majority of people now lived. They had a broader-based livelihood than before: they continued to pursue the agricultural methods they had developed in the Younger Dryas out of sheer necessity, but they could supplement their diet with nuts and game from the forest, which had returned in the new climate. And perhaps they would have continued to live this way, without the need for any major changes, if a new mini ice age hadn't set in, around 6200 BCE.

Once again it was the cryosphere in the north, and in particular the great Laurentide ice sheet in North America, that shifted. This glacier, the world's largest, had now melted so much that it collapsed entirely, driving large volumes of meltwater out into the North Atlantic. This caused the Gulf Stream to stop again, this time for a briefer period of "just" four hundred years, but with results at least as dramatic as on the previous occasion. The problem now was that agriculture had already been discovered while the human population had become much larger. So what were they to do when the climate suddenly worsened and drought took hold?

For many, there was no solution. Others coped by giving up agriculture and opting for pastoralism instead: following flocks of domesticated sheep or goats. Those that could moved to places where there was still fresh water, near rivers and lakes. And, as when animals flock to the water hole in times of drought, this also led to population concentration and people had to learn to live with each other, even with those who weren't family members. Such concentrations would not have been possible while people were living as hunters and gatherers, but since they had now learned to cultivate the land, it had become feasible to live this way. But it demanded social and cultural adaptation, and one "social technology" that now emerged was organized religion. Not shamanism, which is the typical spiritual practice of hunter-gatherers, but a more hierarchical religion. The family-based morality characteristic of the hunter societies was no longer enough. If large numbers of people were to live together, extensive rules were needed to regulate behavior.[71]

In the Middle East, the area that archaeologists and others have studied most thoroughly and where Western civilization originated, there are some striking coincidences. The major steps toward civilization occur on the occasions when the cryosphere in the north is tightening the screws, creating climate crises farther south. First with the Younger Dryas, which kick-started the agricultural revolution, then with the "mini ice age" from 6200 to 5800 BCE, when people moved into concentrated populations, which eventually developed into cities. On both occasions people had to do this because the climate left them with no other options: they were simply forced to adapt. And more such events would come: episodes of drought that concentrated the population around cities and irrigation systems, instigating the development of more centralized political and religious structures. The climate changes forced us to constantly develop civilization.

We know the pattern from biological evolution, from the way scientists such as Elisabeth Vrba have described the consequences of the periods of drought in Africa: people had to adapt or die. But this time, in the Holocene, humans didn't adapt by evolving into a new species; that would have required more time and would therefore scarcely have been feasible. Instead, a totally new type of evolution emerged, a cultural or social evolution in which humans adapted by altering their behavior: making use of new technology (agricultural technology) and new patterns of settlement (living together in villages and towns). This could happen because *Homo sapiens* had acquired a brain that enabled not only finding new solutions but also living together in larger social groups. The Neanderthals had failed to do this and therefore perished.

This new existence was no bed of roses, for either farmers or new city-dwellers. The only thing that increased was the number of people because this way of organizing food production made bigger populations possible. But overall, these people had a worse life than their hunter-gatherer forebears. All the available archaeological traces, including the skeletons, bear witness to this. The farmers were shorter, had worse teeth, and lived a more monotonous life than the hunter-gatherers.

And they were poor. Whereas the hunter-gatherer societies have been called "the original affluent society" (by the anthropologist Marshall Sahlins),[72] because people had what they needed on the whole and rarely worked more than four to five hours a day, the farmers had a much tougher time of it. They literally worked themselves to death and had few pleasures. And although the ruling classes acquired considerable wealth, little benefit accrued to ordinary people. Economic historians such as Gregory Clark and Ian Morris have shown that the lives of most people did not improve for several thousand years, right up until 1800.[73]

As Clark notes, "the average person in the world of 1800 was no better off than the average person of 100,000 BC. Indeed in 1800 the bulk of the world's population was poorer than their remote ancestors."[74] An example of this: "Life expectancy was no higher in 1800 than for hunter-gatherers: thirty to thirty-five years. Stature, a measure both of the quality of diet and of children's exposure to disease, was higher in the Stone Age than in 1800."[75]

Of course, things went up and down, depending on good and bad years for agriculture, but if we look at the slightly bigger picture, there was no increase in affluence for thousands of years, from the agricultural revolution to Renaissance Europe. Only then did living conditions begin to improve, little by little, although only in small pockets. Many explanations have been proposed for the sudden, dramatic improvement that started around 1800 after so many thousands of years. Steam engines were, of course, decisive, as were increases in trade and the cash economy. But here, too, the cryosphere may have played a role, through what has been called the Little Ice Age.

17

WHEN THE ICE RETURNED

This glacier, more to the point, is a tremendously high iceberg, which lies up above the lofty mountains and has shot down on the western side in both Krondalen and Milvirsdalen where, as the valleys spread out, it has extended further and further in its growth. It is sky blue in color and as hard as the hardest stone ever could be, with great fissures and deep caves and chasms everywhere, all the way down into the abyss whose depth nobody, although they have tried to measure it, can discover. When, at certain times, it shoots forward, a great sound is to be heard, as if from organ pipes; and it takes with it, pressed up from the abyss, great quantities of earth, gravel, and stone, much larger than any house could be, which, moreover, it crushes small as sand.[76]

T HUS DID MATTHIAS FOSS, the priest in Jostedalen, western Norway, in 1741, describe how an arm of the Jostedal Glacier caused havoc in his parish. The glacier had crept down into the valley where it advanced in certain years, destroying houses and otherwise making living conditions especially difficult. "And as it brings with it a tremendous cold in summer the nearby land therefore suffers great damage and not just in respect to the crops; moreover the people who cut the hay in the vicinity cannot go about their business at the height of summer, in the strongest heat and sunshine, unless they are sharply dressed for winter."[77]

Valuable land was also destroyed, not least because the glacier water created a "tremendously big river, soon half like a sea in its travel, with hissing and great waves, yes. Given the great power the water possesses, not only are the flat roads, which it overflows, swept away, but the earth is also undermined, so that pastures and great trees are brought down and crushed to pieces and carried to the sea by the river, which eventually causes vast damage, alas—Justedalen's total destruction if hereafter the water continues thus to sweep away and overflow, as it has hitherto, causing much damage to most fields."[78]

It is, remarkably enough, one of the few Norwegian accounts describing the Little Ice Age, a period of cold that lasted from around 1300 to 1850 and had major ripple effects across the northern hemisphere. It wasn't just the fact that glaciers were growing and causing damage in northern Europe and the Alps (the glaciers also expanded in North America, although there is little documentation to be found about what damage they caused there). Rivers that today run free the whole year round froze in winter. This happened to the Thames in England, for example, where winter festivals were arranged on the river in the period between 1607 and 1814. And stretches of sea that people had never seen freeze before now did so: in winter 1837/38, the Skagerrak Strait was covered in ice all the way from Norway to Denmark. In 1622, the Golden Horn and parts of the Bosporus in Turkey froze. In North America, too, it happened: in 1780, the Upper Bay in New York froze, enabling people to walk from Manhattan to Staten Island.

These events were not just curiosities. In at least one case, they had major political consequences and altered national borders. The fact that the map of Norway looks the way it does today can be directly attributed to the Little Ice Age. In autumn 1657, a Swedish army was heading toward Denmark from the south. After the Thirty Years' War, Sweden had conquered territory in today's Poland and northern Germany, but it still viewed Denmark as its archenemy and had long been in conflict with it. While Sweden had a superior army, the Danish were similarly superior at sea and since much of Denmark—not least Copenhagen—was protected by the sea, they didn't feel especially threatened by the Swedes.

But they hadn't taken the climate into account. In the winter of 1657/58, it turned unusually cold, the way certain winters were in this period, and the Danish inland sea began to freeze. In January 1658, the cold was so bad that the Swedish king, Charles X Gustav, took a chance on sending his army across the Little Belt Strait, from Jutland to Funen. It went *almost* perfectly: two cavalry squadrons, men and horses alike, went through the ice and drowned, but most of the army made it across and sent the little Danish army fleeing. Things went even better at the next strait, the much larger Great Belt, and the Swedish army reached Zealand relatively unscathed. Eager to avoid the capture of Copenhagen, the Danes sued for peace with the Swedes. In the Treaty of Roskilde, Denmark gave up a third of its territory, including the Norwegian territories of Bohuslän and Trøndelag. Trøndelag was later returned to Denmark–Norway, but Bohuslän was not and is still Swedish to this day.[79]

More serious for ordinary people in the whole of northern Europe, however, was the fact that the Little Ice Age brought a lot of bad weather with it, thereby worsening crop-growing conditions. The first cold, wet period hit beginning in summer 1315, a period later known as the Great Famine. By the time the climate improved at last in 1322, the bad weather had led to crop failure and famine, taking the lives of more than 1.5 million people in Europe, which didn't have very many inhabitants in those days. In some places, the population was halved. Other such bad years also came later: in France in 1693–1694, Norway in 1695–1696, and Sweden in 1697–1698. In all three countries a tenth of the population lost their lives as a result of famine, and a third of the population in Finland suffered the same fate. In some places, people resorted to cannibalism, and even the king of England had to get by without bread on one journey.

In eastern North America, in the four hundred good years from 800 to 1200, Indigenous societies had been farming corn, beans, and squash. They also lived in cities as large as Cahokia, with some 20,000 inhabitants. These societies experienced massive disruption and depopulation when the

Little Ice Age made their way of life unsustainable. Cahokia was abandoned, and a wave of migration resulted in violent invasion of smaller settlements to the west.[80]

The Little Ice Age didn't just reduce the population in many places; it also destroyed an entire society: the Norse colony on West Greenland, which was established there during a warm period in the Middle Ages and at one point consisted of around 5,000 people. In the 1400s, agricultural conditions became steadily worse, and the remains of the Norse colonists show that their health became poorer. By the end of the century, there were no Norse left.[81]

In Iceland, too, there were periods when the population was threatened with extinction. One reason was volcanic eruptions that worsened growing conditions, but almost as decisive was the fact that a vital resource, cod, didn't like the colder temperatures either and moved south. Cod does best at sea temperatures of between 39 and 45 degrees Fahrenheit. So it didn't just abandon Greenland and Iceland but the Norwegian coast, too, whereas the Dutch now enjoyed good catches.

But the sea banks off the coast of Canada were the places that now benefited most from the cod. This prompted European fishermen to set out on the long voyage across the Atlantic to share in the fantastic fishing. It looks as if Basque fishermen had already been there by the time Columbus came to America. The cod was salted and dried and taken home to Europe, where the Catholics in particular made good use of it: since this was fish and not meat, it could be eaten on Fridays and during Lent.

While some societies disappeared and others suffered declines in both affluence and numbers, some adapted well.

As James Daschuk notes, Indigenous groups in New York State and southern Quebec "came together after several generations of conflict and privation to form the League of the Iroquois, a sophisticated system of governance and diplomacy that continues to function today."[82] Others actually profited from the Little Ice Age. One example was the Netherlands, which enjoyed its best ever period in just that era, what has been termed its Golden Age. This word is usually applied to Dutch painting, but the Netherlands in general achieved major advances: it was, for a time, the world leader in science and technology; and for a time it was also the world's largest colonial power, mostly thanks to its shipping and weapons technology. One reason for this was that climate change led to rich fishing in the North Sea, but it also had to do with the fact that the Netherlands had, at that time, a form of society and culture that was better placed than others to adapt more easily to new times.

And these new times were, to a great extent, the work of the Little Ice Age. Crop failures and the crisis at the beginning of the 1300s (which started long before the Black Death and was exacerbated by it) caused the old feudal society to collapse, and new social structures sprang up. The cities became more important and trade increased. New ships, and technologies such as compasses and clocks, made it possible to travel out into the world in search of new riches. The power of the church was weakened, not least through the Reformation—in part a response to the bad times, which made the greed of the church even more intolerable. But perhaps the most important social change took place in England, where agriculture switched from self-sufficient farming to more specialized production: cash crops, which supplied the cities with food. This agricultural revolution was a prerequisite for the Industrial Revolution, which came in its wake in the mid-1700s and which would totally change the world.

The Industrial Revolution was triggered by the discovery of more effective means of extracting and exploiting fossil energy, initially coal. Most of this coal came into existence during the Carboniferous and Permian periods, which started out warm with large tropical rain forests, but then swung to extreme cold. The period from 340 to 260 million years ago was especially cold, with several cycles in which the ice expanded and contracted. The Earth's continents were different from the way they are today and most of them formed part of one large continent, Pangaea. Much of it lay around the South Pole and this was yet another reason why major ice sheets formed here, extending up to 40 degrees south.

As its name suggests, the Carboniferous period was the time when the Earth's fossil carbon reserves were formed. This came about because of the particular vegetation and animal life in that era, as well as the fact that sea level fell and rose cyclically, causing large amounts of organic material to be submerged and then stored in ways that transformed the carbon (a vital component of living organisms) into coal. Until around 1800, this coal was only used in modest quantities. But when the Industrial Revolution came, with its new means of extracting, transporting, and exploiting coal, usage exploded. The energy that had lain in the ground for 300 million years was now put to use, at an accelerating tempo.

In other words, the Industrial Revolution was based on an energy source that the cryosphere had planted in the earth

300 million years before, during an ice age period. And it was triggered by a far from negligible variation in the climate, which forced through new means of production in precisely those countries that were feeling the pressure of the cryosphere again during the Little Ice Age, in the form of colder weather, poorer growing conditions, frozen waterways, and snow-covered landscapes—so familiar to us from the Dutch painters of those times.

There is still uncertainty about which mechanisms unleashed this period of cold, which was not cold all the time, but involved great fluctuations. Some scientists believe the Milanković cycles (the ice age cycles) may have started it all off; others point to changes in sunspot activity and a decline in thermal radiation from the sun.[83] However, the important thing to remember is that the average temperature did not fall more than 1.8 degrees Fahrenheit (1 degree Celsius) in many places. This is a trifling change by comparison with earlier fluctuations, such as the Younger Dryas and the ice ages—and also in relation to today's much-discussed 2-degree (Celsius) limit. But this single degree would lead to a series of bad crop years and even revolutions (to a certain extent, even the French Revolution can be ascribed to crop failures in the mid- to late 1700s).

One reason why such a small change could have such large consequences is that northern regions are extremely sensitive to climate changes because of the ocean currents—especially when the temperature fluctuates around the freezing point. This is not a change in degree but a phase change. Water in solid form is quite different from water in liquid form. Snow is quite different from rain. And for crops in particular, the difference between a temperature above and below zero means life or death. One degree (Celsius) colder means longer winters and shorter growing seasons. An additional factor was that the highly unpredictable weather in those times was an extra source of problems. People never knew what the next year would be like.

However, the way out of the tough times was the Industrial Revolution, which gave humanity an unprecedented increase in affluence—and population growth. The most important reason for this rise in affluence was the fossil energy the cryosphere had stored 300 million years before, and which we had now learned to use. And for the past two centuries, we have lived well off it, better than at any time in history. But now we see that the use of fossil fuels (not just coal, but eventually also oil and gas) is precisely what threatens to melt the remains of the cryosphere. Considering that it has given us two hundred years of extraordinary affluence by historical standards, that's a fine way to show our gratitude!

18

THIN ICE: WHAT'S HAPPENING TO THE CRYOSPHERE?

One October day the temperature drops 50 degrees in four hours, and the sea is as motionless as a mirror. It's waiting to reflect a wonder of creation. The clouds and the sea glide together in a curtain of heavy gray silk. The water grows viscous and tinged with pink, like a liqueur of wild berries. A blue fog of frost smoke detaches itself from the surface of the water and drifts across the mirror. Then the water solidifies. Up out of the dark sea, the cold now pulls up a rose garden, a white blanket of ice blossoms formed from salt and frozen drops of water.[84]

THUS **PETER HØEG** describes the way sea ice forms off Greenland, through the eyes of Smilla. This is the *new* sea ice that forms every winter in addition to the ice that has survived since the previous summer, and possibly over many summers. Altogether, the new and old sea ice covers almost the entire Arctic Ocean in winter, and a large stretch of it in summer. In recent years, this ice has shrunk dramatically, not just in surface area but in thickness, too. According to the latest report from the Arctic Monitoring and Assessment Programme (part of the Arctic Council), the thickness of summer ice has diminished 65 percent since 1975, and it will have disappeared entirely by 2030.[85] For some—those intending to explore for minerals and oil up here, and those who see the Arctic Ocean as a new and faster sailing link between Europe and Asia—this is good news. But for those who are concerned about climate developments, these are danger signals.

What is happening to the ice in the north on both land and sea has attracted a lot of attention in recent years. In November 2016, the sea ice cover was 1.5 million square miles smaller than normal. In March 2017, too, the ice measurements were at a record low. And the ice isn't just shrinking in area but is also becoming thinner and less constant. "We have an ice system that is totally different from what it was twenty years ago," says Harald Steen, head of the Norwegian Polar Institute's Centre for Ice, Climate and Ecosystems. "The different processes, such as melting, thawing, and ice drift, have taken a turn for the worse. There is less ice. It drifts faster. There is more snow on it. The ice is melting more quickly and there are more open channels through it."[86]

The shrinkage of the sea ice is clearly visible around Svalbard, in the Arctic Ocean north of Norway, where the sea temperature has climbed rapidly in recent years. Ice barely appears on the west coast anymore, other than that calved by the glaciers. This was already noticeable after the beginning of the century, when I visited Ny-Ålesund, nearly 79 degrees north, which is deemed to be the world's most northerly "town." This place was the starting point for several expeditions to the North Pole, by Roald Amundsen and Umberto Nobile, among others. The post to which they moored their airships still stands there. This was a mining town until a serious accident in 1962, which triggered a crisis in the Norwegian government, ending the era of three-time prime minister Einar Gerhardsen; it also inspired a feature film. Now, though, Ny-Ålesund is occupied by a community of international scientists, around thirty of whom spend the winter there, and whose most important pastime in the winter darkness is indoor bandy, a kind of indoor hockey. A lot more people are there in the summer: hundreds of scientists from all over the world. The place is run by Kings Bay AS, the former mining company, which now organizes things for the scientists there.

From the head of Kongsfjorden, where Ny-Ålesund lies, the scientists have been able to watch the Krone Glacier growing thinner and thinner, at an average rate of a few feet a year. Lately, it has also begun to retreat at a rapid pace. When I was there, this retreat had led to the discovery of a new island in the fjord. And the sea ice that previously covered the fjord has vanished. Svalbard is losing ice on both land and water, and it is in the process of becoming more like mainland Norway:

surrounded by ice-free sea and with ever-shrinking mountain glaciers on land.

When people talk about Arctic ice, they are referring to two types. The sea ice is formed by seawater that freezes, and it shrinks and grows according to an annual cycle. This is the ice that is most frequently discussed, because it has diminished dramatically in recent years, both in surface area and thickness. The expectation is that the Arctic Ocean will soon be pretty much ice-free in the summer.

The other type of ice, which has covered roughly the same area from many thousands of years ago until recently, is the glaciers on Svalbard and other islands, as well as the Greenland ice sheet, where Nansen conducted his first expedition. The ice sheet is a couple of miles thick and more or less unchanging from season to season. It has also been considered relatively stable over the longer term. Until recently, scientists had expected it to stay roughly as it is for several thousand more years.

But many recent observations indicate that the Greenland ice sheet appears to be melting much more rapidly than assumed, as are large glaciers. The Jakobshavn Glacier in West Greenland—believed to have calved the iceberg that sank the *Titanic*—is running out into the ocean at a faster rate than ever seen in previous measurements: up to 150 feet a day or 10 miles a year, according to scientists. One reason for this is that warmer seawater is melting the glacier at the mouth (glaciers are like rivers—they just flow more slowly), thereby causing it to slip out more rapidly. But it is also melting from above, as meltwater gathers on the top, creating rivers that seep into cracks in the ice. This is the result of higher atmospheric temperatures.[87] Comparisons with old aerial photographs show that the melting is already visible along the outer edges of this mighty ice cap. Greenland is melting, and it is happening much faster than anybody could have imagined.

This may have dramatic consequences long before there is any notable rise in sea level. When the ice melts, large amounts of fresh water flow out into the sea, disrupting ocean currents in the North Atlantic. What we call the Gulf Stream, although it is actually a branch of the North Atlantic Current, is kept going by a pattern of circulation caused by warm water from the south gradually becoming saltier and heavier on its way north, then sinking to the bottom before returning southward far below the surface. But if large amounts of fresh water come from Greenland or other places in the Arctic, this thermohaline circulation may be slowed and eventually come to a halt.

This is not a hypothesis, but something that has happened several times in history, as I've noted, and has resulted in heat transport to the north being weakened or cut entirely. The Gulf Stream has acted like a gigantic heating cable for us up in northern Europe. This is what has made it possible to survive here and even to work the land at latitudes that are too cold for this kind of thing in other places—Alaska, northern Canada, Siberia.

The most striking example I have seen of this was when I visited the Russian mining town of Barentsburg on Spitsbergen, in the Svalbard archipelago, which is much farther north than Alaska. To my astonishment, I saw that the Russians kept cows outside there and got milk from them. It may be that Russian cows are extremely hardy, but they looked pretty normal to me, without furry pelts. And there they were, all the same, grazing at 78 degrees north—thanks to the Gulf Stream. If it were "switched off" this would scarcely be possible even in Norway.

This is another scenario people assumed lay far in the future. But things can move fast in the Arctic because there are so many tipping points up here—thresholds beyond which developments suddenly become self-reinforcing and irreversible. If we study climate history using the many ingenious methods that have been developed for the purpose, we can see that a series of dramatic, rapid climate changes have happened here.

One person who knows a lot about this is Eystein Jansen of the Bjerknes Centre for Climate Research in Bergen, Norway. The center, which is affiliated with the University of Bergen, is named after Vilhelm Bjerknes, who founded modern meteorology in Bergen in the years during and after World War I. Bjerknes and his collaborators were the people who developed many of the terms and tools that meteorologists and climate scientists still use today. Some of the terminology, like *front* ("warm front," "cold front"), was clearly inspired by the ongoing war, because things can get rough in the atmosphere when cold and warm layers of air collide.

In Bergen, I visit Jansen, who taught me a great deal when I made a TV program about the Gulf Stream some years back.[88] Among other things, he took me along to a climate science seminar in Spain, where I had a thorough introduction to the complexity of this topic, which involves the integration of many different disciplines. Climate research embraces everything from geology (Jansen is a geologist), glaciology (glacier research), geophysics, and data modeling to botany (climate historians use botanical traces).

Jansen also showed me how scientists analyze sediment recovered through core drilling. In a sample from Kråkene-svatnet, a lake in western Norway, we could see clear traces of the Younger Dryas episode 12,800 years ago. Jansen later became involved in the work of the UN's IPCC and has been one of the main authors of the reports it issues.

Jansen has done research into earlier climate changes, which have often been abrupt, involving temperature shifts of up to 18–27 degrees Fahrenheit in the course of a decade. This is something we have been spared in the abnormally stable period we call the Holocene, the past 11,600 years. As a result, we have gotten used to thinking that this is how it always is: that the climate changes gradually, in a linear fashion; the way we tend to envisage climate change later in this century, with a rise of perhaps a couple of degrees (Celsius)—we hope. It doesn't sound all that bad and that may be one reason why we haven't totally come to grips with the climate challenge.

Jansen is one of our foremost experts in the interpretation of sediment cores from the ocean bed, where, for example, the remains of small animals can tell us what the climate was like in different periods. We know that certain insects reproduced at some temperatures and not at others. By drilling far down into the ocean bed and sticking a tube down there, we can extract drill cores that can be read off, like a card index of the past climate. A similar technique can be used in ice. Danish scientists, for example, have spent several decades analyzing drill cores from the Greenland ice, where it is also possible to track climate history tens of thousands of years back by studying, say, dust (which tells you where the wind came from) and the contents of air bubbles in the ice (which can tell you something about carbon dioxide content and temperature).

Now Danish scientists, including some from the Niels Bohr Institute, a world leader in ice-core research, are working with Jansen and other scientists from the Bjerknes Centre. They are well underway with a relatively large research project, ice2ice, financed through US$14 million from the European Research Council's prestigious Synergy Grants. The aim of the project is to understand what caused the many dramatic and rapid climate changes during the ice age that have been proven by both ice-core scientists and paleo-oceanographers (scientists studying the history of the oceans in the geologic past). The name ice2ice indicates the project's working hypothesis: that the development of ice in the Arctic—both the sea ice and the Greenland ice sheet—is the key to these changes.

The last ice age was not, in fact, a stable cold period: there were a number of dramatic shifts, which have been difficult to explain until now. In particular, the ice2ice project aims to cast light on the so-called Dansgaard–Oeschger events (named after the scientists who discovered them). These events involved rapid periods of warming followed by even more rapid cooling, which occurred at intervals of around 1,500 years. Scientists see certain parallels with the situation today, and therefore think the knowledge they can obtain about these events can give us better models for predicting future climate developments.

One of the keys is the albedo effect. As I've said, snow-covered sea ice can reflect up to 90 percent of solar radiation, in contrast to open sea, which absorbs most of the radiation, and with it the heat. Once we grasp that the sea ice in the northern hemisphere can cover an area as large as the whole of Antarctica, we realize that this has great significance for the temperature. Less ice and snow lead to more heat, in a self-reinforcing process.

Another reason why the sea ice affects the climate is that it places a "lid" on the sea, which prevents heat exchange between sea and atmosphere. The seawater may be warm beneath the ice but this does not affect the atmospheric temperature. But it is even more complex than that. What keeps the ocean circulation going, making northwestern Europe so much warmer than other places on the same latitude, is thermohaline circulation, as I've explained.

This process can be interrupted. When the Gulf Stream stopped during what is known as the 8.2 ka event (8,200 years ago), it was because masses of fresh water flowed out of North America and settled on top of the salt water. That brought ice age temperatures back to the north. But why did such abrupt shifts happen during the ice age? And could something like this happen again in our times?

It could, for several reasons. More fresh water may come and settle on top of the seawater if the flow from the Greenland ice, other glaciers, and Arctic rivers increases, hastened by the warming. It may also happen because the sea temperature rises, ultimately disturbing the stable stratification currently typical of the sea up here.

There are also other elements of uncertainty, one of which is how the jet streams are affected. What we have seen in recent years is that the polar jet stream, a wave of air currents on top of the troposphere (the lowest layer of the atmosphere) in the area where cold air from the Arctic meets warmer air from the south, is moving in larger arcs and extending farther in both southerly and northerly directions. This has led to abnormally high temperatures in Svalbard and abnormally low temperatures elsewhere; it has also brought generally more extreme weather, leading to weather records in the areas of temperature, precipitation, and wind. There is also a danger that higher sea temperatures will hasten the ongoing melting of the Greenland ice. This may happen because warmer sea will melt the ice where it meets the ocean (glaciers are constantly moving out toward the sea), speeding up the drain-off—a phenomenon that has already been observed.

The climate in the Arctic is complex and can rapidly undergo drastic change. At some point, the warming may reverse, triggering a cooling, as has been seen on several occasions. This is something Jansen and his colleagues hope they can say more about once they have carried out the ice2ice project. It is uncertain whether they will find out exactly when the tipping point will occur. But what research shows is that we cannot take anything for granted in the Arctic. The Kingdom of Frost may strike back, faster and harder than we believe.

This is how things may look in the north, at any rate. In the southern part of the cryosphere, the Antarctic, people have previously believed that the ice was more stable and that it would be thousands of years before any noticeable changes could happen there.

But disturbing reports have recently come from the land around the South Pole, too. In particular, the ice that projects out into the sea—"ice shelves"—from West Antarctica and the Antarctic Peninsula has shown signs of weakness. The most frequently cited are three of the Larsen ice shelves along the eastern coast of the Antarctic Peninsula. The ice shelves are named after the Norwegian sea captain Carl Anton Larsen, who sailed the whaling boat *Jason* as far south as 68 degrees 10 minutes in 1893. From north to south the three ice shelves are called Larsen A (the smallest), Larsen B, and Larsen C (the largest). And it is also from north to south that these ice shelves have begun to break up and partially collapse.

It began when Larsen A disintegrated in January 1995. Larsen B was next, collapsing in February 2002. And now the largest one, Larsen C, has started to crack and is beginning to disintegrate.[89] This shelf covers an area of approximately 19,000 square miles. Whereas people previously thought this ice shelf was stable, observations in 2016 showed a growing crack, which was over 68 miles long at that point. In 2017, a massive iceberg about the size of Delaware broke off from the shelf. As of July 2018, the iceberg had traveled 28 miles northeast before getting stuck behind the Bawden ice rise, a kind of scaffold for Larsen C. If that ice rise is destabilized, the rest of the shelf could collapse.[90]

In itself, this separation will not have so many consequences: this is, after all, ice that is already floating on the sea, so it won't make the ocean rise. But when this shelf breaks off, the remaining ice will be vulnerable to further collapse. If the whole ice shelf vanishes, this will mean that the mainland ice, currently held in place by the shelf, will be able to run out unhindered into the sea. Other places on the edge of the Antarctic Peninsula and West Antarctica have also been showing signs of instability recently. In 2008, a large ice floe separated from the Wilkins ice shelf on the western side of the Antarctic

Peninsula. In October 2016, NASA readings showed that three glaciers on West Antarctica—the Smith, Pope, and Kohler Glaciers—have begun to melt.[91]

On the other hand, East Antarctica, the place that contains the most ice not just in Antarctica but the entire globe, has seemed more stable. Scientists had spotted few signs of melting there, in contrast to the more unstable but smaller West Antarctica. But now, Australian scientists from the University of Tasmania in collaboration with American scientists from the University of Texas have observed unexpected conditions in and around the ice shelf that fronts the largest glacier on East Antarctica, the Totten Glacier. Here, too, the ice shelf is holding the rest of the glacier in place and preventing it from flowing out into the sea. This ice shelf is now melting at a rate of between 69 and 88 million tons a year, which means that it is losing 33 feet of thickness every year.[92]

These scientists also found out why it is happening: warm seawater far below is flowing in beneath the ice shelf and melting it from underneath. Until now, Antarctica has been sheltered by cold sea currents around the continent, but this appears to have changed lately. In fact, these changes are what have made it possible to get close enough to the Totten Glacier to take these readings.

When the ice shelf has gone, the world's biggest glacier will begin to slide out into the sea, potentially causing sea level to rise 11.5 feet from this glacier alone. It is uncertain how long it will take. This may have been what happened around 3 million years ago, when sea level rose some 10 feet over a short period—even though the mean global temperature was only slightly higher than now. In that warm period, the temperature was just 3.6 degrees Fahrenheit (2 degrees Celsius) above today's levels, within the so-called 2-degree limit, and yet this led to a full 72-foot rise in sea level.

In both the Arctic and the Antarctic, scientists have seen things changing much faster than they had expected. And they have discovered mechanisms in the geophysical system that may become self-reinforcing and may accelerate. When the ice melts, this affects the albedo, ocean currents, and air currents as well. And the people who are studying the climate of the past know that rapid, dramatic climate changes have happened many times before.

Perhaps just such a climate change is already underway. The climate systems are so complex that we cannot say anything for sure, but the most likely prospect appears to be that the cryosphere will continue to shrink, at least for some time into the future. If this happens, what will it mean for the environment and for people around the globe? As we have seen, reindeer-herding Sami and other Arctic people are not the only ones who rely on the cryosphere. The people who live in the vicinity of the cryosphere's "outposts"—such as the mountain ranges of Asia and America—will be among those who suffer the brunt of it if the ice and snow disappear.

— 19 —

THE ROOF
OF THE WORLD
IS MELTING

F EW NATURAL PHENOMENA have a greater power of attraction than mountain glaciers. Some of Norway's most important tourist attractions are offshoots from the Jostedal Glacier: the Briksdal Glacier and the Nigard Glacier. People come all the way from Asia in their thousands to see them, maybe to walk on them a bit if conditions allow it. And Norwegians make the pilgrimage to the Hardanger Glacier to take part in the Constitution Day parade there on May 17. Glaciers simultaneously represent beauty and danger, what Kant called "the sublime." They have been there for thousands of years (or so we think, although some are actually much younger) and are the remnants of the great ice cap that covered the whole of northern Europe and much of North America a full 20,000 years ago, with a thickness of well over a mile.

The glaciers entice and the tourists are keen to get as close as possible. *Too* close, it often turns out. Almost every year we hear about accidents where tourists have been hit by ice blocks. They may have got that selfie they were after, but it was also the last thing they got. The glaciers are actually "alive": they are like rivers, except that they move much more slowly, according to a different clock.

The fact that glaciers can be unstable and dangerous is extremely well known to the Sherpas—the Nepalese mountain people who almost carry rich tourists (and at the very least their luggage) up to the peaks of the Himalayas. In April 2014, an ice avalanche took the lives of sixteen Sherpas at Everest Base Camp, and these kinds of avalanches are not uncommon. But when there were avalanches at two neighboring glaciers in western Tibet in 2016—one in July and the other in September—that was more unexpected. The fact is that western Tibet has been considered a stable area when it comes to glaciers and, unlike in other parts of Tibet and the Himalayas, people haven't noticed the glaciers here melting. On the contrary, the impression was that the glaciers here were growing; that was the way it looked from satellites, at any rate.

But some of the most important things that happen to glaciers aren't visible from above because they take place inside the glaciers and below them. That is what Lonnie Thompson of Ohio State University, the world's most experienced and renowned mountain glacier researcher, says. Along with

Chinese colleagues, he was invited to Tibet to discover the causes of the first of these glacier collapses, which took place at the Aru Glacier on July 16, 2016, killing nine yak herders down in the valley. This glacier had been stable for several decades. The scientists' conclusion, which was published in the *Journal of Glaciology* in February 2017, was that global warming was the most probable cause.[93]

Scientists tend to be cautious about linking concrete events to the phenomenon of global warming, so how could Thompson and his colleagues draw this conclusion? Because of the way the avalanche happened: ice masses as large as this—77 million tons—could only collapse as rapidly as they did, in the course of four to five minutes, and cover such a large area if the ice had been "lubricated" from below by meltwater. In order for such a large mass of ice to be lubricated in this way, a lot of ice must have melted. Temperature readings also show that the temperature in the area has risen by an average of more than 2.7 degrees Fahrenheit (1.5 degrees Celsius) over the past fifty years. A sign that this event was not random came later, in September, when another nearby glacier collapsed in the same way. This indicates that the events had a common cause, and warming is the most probable culprit.

Thompson has plenty of empirical material to build on, and had already carried out a similar, comprehensive study of the glaciers in the world's "third pole"—the Himalayas and Tibet, as well as the more westerly Karakoram and Pamir Mountains—together with Chinese colleagues. The study, which was led by Tandong Yao of the Chinese Academy of Sciences in Beijing, was the most comprehensive study hitherto conducted into the glaciers up here on the Roof of the World.[94]

The status of the glaciers in this area has been hotly debated. Part of the problem is that there were no decent data to build on. Western glaciologists haven't had access to the region until recently, so they do not know what the glaciers' "normal" annual development (mass balance) looks like; nor do they know how far the glaciers have previously extended or what their thickness has been. The local scientists had neither good enough methods nor good enough equipment to take readings that could be used for comparative purposes. As a result the forecasts for the development of ice on the Roof of the World have typically been uncertain, and the messages have been somewhat conflicting. This has made it easier to sweep the gloomy prospects under the carpet. One example that is constantly brought up is the mistake in the 2007 report by the UN's IPCC, which forecast that the glaciers here would disappear by 2035. It should have said 2350! The basis for the figures was too imprecise anyway, but people are now exploiting this mistake for all it's worth. It doesn't help that subsequent research has reached quite different conclusions.

One example is a study using satellite data from GRACE (Gravity Recovery and Climate Experiment), which indicated that the glaciers are not shrinking as rapidly as had been assumed.[95] But there were a number of problems with this study, according to Thompson. First, it ran for too few years, only seven, whereas climate scientists prefer to use time series with a minimum of thirty years so that they can distinguish between weather (which changes drastically from year to year) and climate (which is a more prolonged trend). Second, the satellite misinterpreted the signals from the ground. As Yao told *Nature* on this issue, after his team's own field observations showed a quite different picture from the satellite data: "As the GRACE satellites can only feel the gravitational pull and can't tell the difference between ice and liquid water, they may have mistaken expanding glacial lakes for increases in glacier mass."[96] After all, one consequence of glaciers melting is precisely that they form such lakes.

In their field studies, Yao and Thompson have reached a totally different conclusion from the GRACE study: after scrutinizing data gathered over thirty years from more than 7,000 glaciers in the Himalayas, Tibetan Plateau, Karakoram, and Pamirs, they found total ice loss of around 9 percent since the 1970s. But this figure conceals large regional variations, linked to the fact that the climate has changed in different ways. On the west side of this mountain region, in the Pamir Mountains, which mostly lie in Tajikistan, there has been an increase in westerly winds, leading to more precipitation. As a result, the glaciers here have not diminished notably.

However, they have done so in more southerly and easterly regions of the Himalayas and Tibetan Plateau. Here, the temperature has risen and many glaciers have shrunk discernably. This doesn't just apply to the surface, but also the interior of the glacier, as Thompson and his colleagues have measured through drilling. One example is the Naimona'nyi Glacier, which provides water to the Indus, the lifeblood of Pakistan. It has shrunk at a rate of 16 feet per year over a period of thirty years. Thompson has carried out drilling in this glacier and discovered that holes have formed at many points in the ice and the meltwater that gathers there causes further melting. "It means that the glaciers are wasting much faster than just the loss of area, but they are also wasting from the top down, which means they are losing ice volume rapidly." He adds: "This is significant for water resources in the Indus River, as it is believed that 40 percent of the water discharge in that basin in the dry season comes from melting glaciers."[97]

Thompson's and Yao's studies indicate that a trend that has been visible for the past 150 years since the end of the Little Ice Age is not only continuing but apparently accelerating. Backing them up, a study conducted jointly by ICIMOD (International Centre for Integrated Mountain Development) and UNEP (United Nations Environment Programme) stated that "the Himalayan glaciers have retreated by approximately a kilometre since the Little Ice Age."[98]

This conclusion is also supported by a recently completed Norwegian-Indian study, GLACINDIA,[99] led by Atle Nesje of the University of Bergen, one of the world's leading scientific experts on mountain glaciers. He grew up close to the Jostedal Glacier in western Norway, and has spent his life studying the dynamics of mountain glaciers: how they expand and contract. The main focus of the project in the northwest of the Himalayas was the Chhota Shigri Glacier. The scientists have followed its mass balance for decades, compared it with meteorological data, and reconstructed the glacial development here during the Holocene (i.e., the past 11,600 years). These are just the kind of data that have been missing and whose lack has made it difficult to produce good forecasts.

The Chhota Shigri faces north and lies partially in the monsoon shadow. That means it receives only a little of the precipitation from the monsoon and, other than that, a small amount from the westerly winds. The meltwater from this glacier ends up in the Indus, which is absolutely crucial to

Pakistan's existence. Scientists have studied its mass balance every year since 1955, measuring how much the glacier has retreated or grown. Since the new millennium, the trend has been for a more drastic reduction: a negative mass balance of 0.6 meters water equivalent per year, which is double the level for the period from 1955 to 1999. Since there is little precipitation here to start off with and precipitation levels have shown little change, the probable reason why the glacier is shrinking is that temperatures have become warmer, with a rise of 3.6 degrees Fahrenheit (2 degrees Celsius) over the past sixty years. The temperature in the third pole is not only rising, but rising at a more rapid rate than the global average.[100] As Lonnie Thompson put it, "The higher the elevation, the greater warming we have. This is in line with the observation that the vast majority of glaciers in Tibet and the Himalayas are retreating."[101]

So it looks as if the Roof of the World is in the process of melting. The glaciers that have been here for thousands of years, serving as a water tower for billions of people in South Asia and China, are shrinking discernably, but we cannot say for sure how long it will take for them to vanish. If the melting continues and ultimately accelerates, the consequences in the first instance will not be a lack of water but the opposite. More floods will come, wreaking greater destruction because there will be more meltwater. And the biggest problem is generally not the water itself, but what it brings with it: ice, stones, gravel, and soil—as in the Tibetan glacier collapse mentioned earlier and ice avalanches like the one at the Everest Base Camp. Dangerous situations have also arisen when ice, stones, and gravel carried down by avalanches have created temporary dams. When, after a time, they are breached, this leads to catastrophes downstream. These "glacial lake outburst floods" (GLOFs) have become a common occurrence in Nepal, which has experienced twenty-one such events in recent times.[102] And in the Himalaya region, two hundred glacier-dammed lakes have now been recorded which are at risk of breach, with potentially catastrophic consequences for those downstream.[103]

Over the long term, on the other hand, when the first phase of melting has ended and the glaciers are spent, the really vast problem will emerge: the lack of water during the dry months, which will affect perhaps as many as 1.5 billion people. For several thousand years, since the end of the last ice age, the glaciers on the Roof of the World have done their job, serving as water towers for some of the world's oldest and most powerful civilizations, from the Indus Valley in the west to China in the east. And they still keep at least a fifth of the Earth's population alive.

But the Roof of the World is not the only place where glaciers serve as vital water sources; it's just that people haven't noticed their significance. One reason is that most people haven't seen them, much less come into contact with them. Another is that very little mapping has been carried out to establish how much water comes from glaciers—and the periglacial environment, the frozen earth in the vicinity of the glaciers. In this respect, we must not only consider the amount of water, but also the fact that the glacial environment provides water in periods of little rain after the melting snow has done its bit.

This applies to all areas of the globe where there are glaciers: Central Asia, Turkey (site of the sources of two of history's most important rivers, the Euphrates and the Tigris), the countries around the Alps, Japan, New Zealand, and even Africa: one of the sources of the Nile is the Rwenzori glaciers in Uganda, for example. Indonesia also has glaciers, as does Australia (on some islands in the Indian Ocean).

But perhaps most of all, this applies to America, both North and South. One example is the desert city of Las Vegas, which takes its name from geological formations that are created by meltwater (vegas), a type of landscape that can also be found in the Andes. It is also the case that as much as 90 percent of all the water Peru uses comes from glaciers (unsurprisingly, the Incas worshipped the mountains for their water), while Bolivia's capital, La Paz, obtains a third of its water from them. Even the inhabitants of Mexico City drink glacial water, although the glaciers are a long way away.

Only in recent years have we become aware of the significance of the mountain glaciers. And it is events in the Americas, in particular, that have contributed to this: the water crisis in California and the Andes, and the conflict between mineral interests and glacier preservation in Argentina and Chile.

— 20 —

INVISIBLE GLACIERS AND CRYOACTIVISTS

"**C**AN YOU SEE the tongue?" Connie Millar asks. I look up at the valley between two mountains and, yes, perhaps I can just about see it, a swelling that does, indeed, resemble a tongue. It is in front of what looks to us lay people like a regular pile of stone and gravel, the kind left by a landslide. But according to Millar, who has been doing research up here for forty years, this "tongue" is one of several indications of something I didn't even know existed: an invisible or rock glacier. "We can also tell from the vegetation," she continues. "Nothing grows up on the rock glacier, but because of the meltwater there's lush vegetation down below. We've been up there and taken readings and we know there's a glacier underneath."

We are up in the Sierra Nevada, California, at an elevation of 10,000 to 13,000 feet, in an area that Connie and her colleague Bob know extremely well. They are scientists employed by the US Forest Service and divide the year between these mountains, where they have a station by a saline soda lake, Mono Lake, and Berkeley, where they spend the winter in their offices like most other scientists. And part of their job is to keep an eye on the glaciers up here.

Glaciers in California? Ice isn't exactly something we associate with this sunny state. Okay, snow may fall on the mountaintops, but glaciers are only found in colder parts, aren't they? And besides, you don't see any glaciers when you drive around here, even if you cross the mountains on the way to Reno or Las Vegas, do you?

No, there isn't all that much to see. It's true there are a few regular glaciers, like Lyell Glacier, but they aren't visible from the usual roads. However, as long as you know how to interpret landscape correctly, you can discover the *invisible* ones, the rock glaciers that we have come to see. These are glaciers that are covered in stone and gravel or largely consist of them, with ice in between the stones. But they can still be active—in other words they can move, retreat, and grow. And they make no small contribution to the water supply in California—a fact that Connie Millar and her colleagues have tried to communicate to the authorities and the public for many years now, with little success.

But there are glaciers like these all over the world: in the Alps, on Svalbard, and, as noted, in California. Even on Mars, scientists have observed what are thought to be rock glaciers. Far from being trivial curiosities, these are vital components of the bigger ecological and hydrological picture. They affect plant and animal life, since some species will disappear if these glaciers melt, and they help provide water for the people below during the dry seasons. In Chile, rock glaciers provide the water supply for the capital, Santiago. And in many other places like California the rock glaciers are more important than people imagine, Millar says. The problem is that almost nobody has heard of them, even though people like Millar have spent several decades trying to educate the population about them.

Even among glaciologists, rock glaciers are so "new" that nobody is quite sure how they come about. The most common theory is apparently that they are formed when ordinary glaciers contract, which causes the stones that were originally dispersed in the glacier to end up lying closer together. They may also be formed from rock avalanche deposits that freeze.

The point is, in any case, that this is a separate type of formation, which is not interchangeable with regular rock avalanche deposits. They have a life of their own, so to speak, and move, although more slowly than ordinary glaciers. They vary depending on the season, melting a bit and then freezing again. And, just like other glaciers, they serve as water reservoirs, sustaining ecosystems, agriculture, and human societies.

I hadn't heard of invisible glaciers until I read Jorge Daniel Taillant's book *Glaciers: The Politics of Ice*.[104] Taillant, who is a political scientist by training but has also studied glaciology, is now head of the Center for Human Rights and Environment and has long been engaged with events in the Andes, including those involving the "invisible" rock glaciers. Many of these lie on the border between Argentina and Chile and play an important role in supplying water to several fertile valleys in both countries. Taillant has made calculations that show how much water even a small rock glacier can yield. It is often enough to sustain a town, providing farmers and wine-growers with the water they need in the dry seasons. But precisely because they cannot be seen, at least not with untrained eyes, the rock glaciers are at risk.

Without anybody realizing it, several glaciers were in the process of being destroyed by construction work linked to mineral exploration. The central Andes are a veritable El Dorado, rich in gold and other precious metals. And here, at elevations of 13,000 to over 16,000 feet, foreign companies had been working on mining projects for several years, doing construction work with dynamite and bulldozers that involved the partial destruction of glaciers.

This had been going on since the 1990s, and it was only in 2006 that Argentina's new environment minister, Romina Picolotti (a former environmental activist who has won the Norwegian Sophie Prize, among others), was made aware of what was going on. She was especially provoked by the activities of the world's largest gold miner, Canadian company Barrick Gold. In the Pascua-Lama mining project, which is located partly in Argentina and partly in Chile, the company had already made major incursions into the glaciers, such as dividing one glacier in three.

It was local people, including environmental activist and glaciologist Juan Pablo Milana, who alerted Picolotti to the issue at a meeting. This prompted her to collaborate with glaciologists and environmental conservationists to launch a totally new project: a law to protect glaciers, including rock glaciers, and—as work progressed—also the periglacial environment, the areas nearest the glaciers. Despite the glaciers' vital importance for the water supply and the environment, especially in the dry part of the year, no other country in the world had yet drafted a law to protect these resources.

Work on the law started off well and it was actually approved by the Argentine parliament on October 22, 2008. Part of the reason for this was probably that few people realized there were glaciers in the areas where mining work was underway, or in areas that might become relevant to such work. People thought the only place you'd find glaciers was down in Patagonia, not in the central Andes. This is understandable enough, since most glaciers here are small and are often either rock glaciers or partly concealed by stones and gravel. That's why people didn't see that this type of law might clash with economic interests.

The resolution hit Barrick Gold like a bolt from the blue, and the company immediately made contact with people in the corridors of power. The end result was that the president, Cristina Fernández de Kirchner, vetoed the law. This prompted Picolotti to resign, an event the political opposition viewed as a good stick to beat the president with. After a bit of to-and-fro, a new, revised version of the law was eventually approved, this time without veto.

It has since emerged, however, that the authorities have largely allowed the mining companies to carry on as before. Things came to a head in winter 2016/17, in part because of several environmental scandals linked to another of Barrick Gold's projects in the central Andes, Veladero. Now, in late

2018, their business has been brought to a halt after Chile's environmental court confirmed an order to close the Chilean side of Barrick's stalled Pascua-Lama project.[105] It is still unclear how this will end. But at any rate these events did a lot to raise awareness of the mountain glaciers and their significance, and they also resulted in the world's first ever law on glacier conservation, among other things. In Chile, too, glaciers have attracted a lot of attention, and although no laws have been passed, as in Argentina, people now have a more conscious relationship with the glaciers there.

This struggle has also given rise to a new brand of environmental activist—the *cryoactivist*. That said, other people have also previously realized the significance of glaciers and attempted to protect them. In Peru, glaciologist Benjamín Morales Arnao experimented with sawdust as a means of protecting glaciers against melting. After just three months, the ice that was covered in sawdust was 16 feet higher than the unprotected ice. It is even possible to cultivate glaciers, as engineer Chewang Norphel of India has demonstrated. In 1987, he became so concerned about the shrinkage of the glaciers that he decided something had to be done. He built small dams to channel the water out of rivers and into fields where it could freeze in the winter, thereby accumulating ice. In this way, he managed to create artificial glaciers, which keep several villages supplied with water in the dry periods.[106]

These are just scattered efforts by individual people. But they demonstrate that glaciers actually *can* be protected. And the cryoactivists in Argentina and Chile have shown that it is possible to combat the outright destruction of glaciers that is taking place.

21

FROZEN EARTH

THE ROCK GLACIERS are not the only "invisible" parts of the cryosphere. The Kingdom of Frost is much larger than it looks from a plane, or in a photograph taken from a spacecraft. Large stretches of land in the northern hemisphere, as well as somewhat smaller areas in the south, consist of what we call permafrost. Many people aren't aware that we have this kind of thing in Norway, but I grew up in an area of permafrost on Finnmarksvidda. Other such areas are found on the Varanger Peninsula in the far north of Norway, and certain mountain areas where the frost helps stabilize the terrain.

Those of us who live up in the north know that soil can freeze in winter. We experience it every spring, for example, when the frost begins to thaw. There's a special term for it: frost heave—the mounds caused by the expansion of water on freezing. When I was growing up in northern Norway, in a time when we had almost nothing but gravel tracks, driving along the roads could be an extreme sport in spring, and it still can be on certain stretches.

But in many places in the Arctic and mountain regions, the soil doesn't thaw completely in the summer. Maybe a thin layer on the very top, what is known as the active layer, but beneath that the earth is frozen and has generally been that way for thousands of years. This is what we call permafrost (the scientific definition requires it to have been frozen for at least two years). In some places, this frost may go very deep indeed, all the way down to the point where it meets the heat of the Earth's core. By the Lena River in Siberia, permafrost has been measured at depths of 4,500 feet.[107]

And it is, in fact, in Siberia that we find the largest expanses of permafrost, although there are also large areas in northern Canada and Alaska, as well as parts of Greenland, Svalbard, Novaya Zemlya, and other Arctic islands that are not covered in ice. The thing is, the ice has an insulating effect and, at the same time, the great pressure creates heat: as a result, glaciers in many places float on earth that is not frozen, which makes them more mobile than they would otherwise be. Other relatively large areas of permafrost lie in mountain regions such as the Tibetan Plateau and beneath the surface of the sea on the continental shelf around the Arctic. This area was above water during the ice age and the permafrost from that time has remained.

Scientists have calculated that permafrost covers as much as 24 percent of the ice-free land area in the northern hemisphere, around 7 million square miles. There is more sea in the southern hemisphere, and only a few ice-free areas in the Antarctic, Patagonia, and the Andes have permafrost. But all told there is nonetheless quite a lot of permafrost in the south, too, and this is a crucial factor for the climate.

What is this permafrost? It may be frozen soil, stone, or sediment. And although the water content is not high—permafrost accounts for only 0.022 percent of the Earth's water reserves—the frost makes a difference. As the frost heave on roads demonstrates, frost can have major implications for stability. And many settlements in Alaska and Siberia are now noticing this: roads, buildings, and other infrastructure are beginning to subside as the permafrost thaws. Melting permafrost also threatens to cause the collapse of large mountainsides in places like Nordnesfjellet in northern Norway, where 390 million cubic feet of rock risks sliding down into the sea, causing a tsunami over 145 feet high that would hit many settlements around the fjord. For now, the permafrost is holding the mountainside in place. The much-discussed landslides that have taken place in Longyearbyen on Svalbard were also caused by thawing permafrost.

But just as importantly, the frost is significant for life itself—there is life even in areas of permafrost—and especially for the microorganisms in the soil. They control a large share of the carbon turnover—which, again, influences the climate. Some tiny microorganisms live at subzero temperatures, but the areas where the upper layer thaws every year, in particular, can have a highly active life because the midnight sun enables photosynthesis to take place night and day. And this doesn't just apply to microorganisms: the tundra and other areas of permafrost have their own flora, especially species of lichen, moss, fungi, and grass—even flowers and shrubs. This was something we saw every summer on Finnmarksvidda: an explosion of flowers, heather, and berries, especially cloudberries.

There is also wildlife that knows how to make the most of all this bounty, like lemmings, ptarmigans, and hares, even bears in some places, not to mention reindeer/caribou, which graze on reindeer lichen in particular. So the permafrost is far from lifeless. On the contrary, Finnmarksvidda, for example, is a productive grazing area for reindeer and a popular place for hunting, especially for ptarmigan, which attract hunters all the way from southern Norway. And, as anybody who has been there in summertime can confirm, there are also enormous numbers of mosquitoes that don't just bite but force their way into your mouth, nose, and anywhere else they can possibly access. Permafrost does not imply a lack of life.

We find most of the permafrost in the tundra, the term for the part of the Arctic where no trees grow. (There is also an alpine tundra, in mountain regions.) Over thousands of years, the grass, heather, and shrubs on the tundra have absorbed carbon from the atmosphere in the form of carbon dioxide, which, together with solar energy and water, powers photosynthesis, the very basis for most life here on Earth. The carbon has gradually accumulated in the soil, where microbes digest organic waste from the plants. This happens in the active part of the permafrost, the layer that thaws each summer. But in winter it freezes again and, over time, large amounts of frozen carbon accumulate, which do not have any impact on the climate as long as they remain frozen.

But we can no longer count on them staying that way.

22

CLIMATE BOMBS
IN THE TUNDRA

BUBBLES ON THE tundra? Alexander Sokolov, an ecologist at the Russian Academy of Sciences, had been doing research in the Yamal Peninsula and the surrounding islands for twenty years, but he'd never seen anything like this. The bubbles, maybe a few feet or so in diameter, were round and protruded from the ground. When you stepped on them, it was like treading on jelly—something that we can easily imagine even though few of us have ever done it.

It was in summer 2016 that Sokolov and his colleague Dorothee Ehrich came across these bubbles when they were doing fieldwork on Bely Island in the Kara Sea, north of Siberia's Yamal Peninsula. The island is a popular area for research, in part because polar bears like to visit. But when the scientists discovered several bubbles, their summer took an unexpected turn. They decided to investigate the contents of the bubbles. When they punctured one of them, it released odorless "air." The next time, they brought instruments they could use to analyze the emissions and found that the air in the bubble contained two hundred times more methane than normal air, and twenty times more carbon dioxide. The bubbles were quite simply climate bombs—if, that is, there were a lot of them, or might be a lot of them in future.[108] And that has proved to be the case: in spring 2017, scientists reported that they had registered 7,000 methane bubbles below the ground in Siberia, bubbles that could burst at any moment.[109]

And the bubbles weren't the only sign that strange things were going on in the tundra: in 2013 came the first report of a loud, mysterious bang on the Taymyr Peninsula, audible some 60 miles away.[110] A while later, the site of the explosion was discovered by some reindeer herders who almost fell into a big crater, which has since expanded.

Similar phenomena have since been observed in other parts of the Arctic, too, such as northwest Canada. In fact, the reports from Russia prompted several of America's national laboratories to start monitoring methane emissions from the North American tundra. What scientists then discovered was that things had also begun to bubble up here, mostly in freshwater pools. And what was bubbling up was pure methane. "A tossed match would have set it ablaze," a scientist told reporters from the Associated Press.[111] The methane bubbles up from sources on the beds of lakes and tarns.

So it manifests itself in many different ways, from jelly-like bubbles in the tundra, to explosions that leave craters, to methane that bubbles up from fresh water, but the underlying cause is the same: the permafrost is thawing. For thousands of years, organic material from plants and animals has lain there peacefully frozen in the permafrost. But temperatures in the Arctic have risen in recent years, climbing more than 4.5 degrees Fahrenheit (2.5 degrees Celsius) since 1970—a much faster rate than elsewhere on the planet. And this has made the tundra start to thaw, along with the organic material, which has then, in turn, served as food for microbes. This is how the breakdown may have started, resulting in emissions of carbon dioxide and methane (where there is water). The precise way this process works depends on the circumstances. In some places, the gases seep up; in others they have to break through layers of ice, thereby causing explosions.

Regardless, though, the end result is greenhouse gases. And the amount of carbon that is stored in the permafrost is dizzying: the upper 10 feet of permafrost alone contains more carbon than is currently present in the atmosphere (and in some places, it may go over a thousand feet deep). According to the research team led by ecosystem ecologist Ted Schuur, 10 to 15 feet of permafrost looks set to thaw over the next hundred years.[112] In addition, emissions may come from the permafrost beneath the ocean bed, where there are also large carbon stores.

However, this is also where one of the solutions to the climate challenge may lie, because large amounts of methane are trapped in small "cages" of ice known as gas hydrates. In principle, these could be used as environmentally friendly fuels, causing very low emissions of greenhouse gases, if only we could find a way to extract and exploit this raw material.

That is something scientists and technologists worldwide are now working on. We do not yet know how it will go; the challenges include preventing unintended emissions. But in any case, this does hold out the prospect of an environmentally friendly replacement for ordinary oil and gas.

The permafrost can release more than carbon (in the form of carbon dioxide) and methane. When the tundra thaws, a lot of strange things come to light. In summer 2016, an epidemic of deadly anthrax broke out in a reindeer flock in Siberia. Over a thousand reindeer died and people were infected as well, with dozens hospitalized; a child died.[113] The most likely cause of this outbreak was frozen cadavers of infected reindeer, which had been lying in the tundra and had begun to thaw. This also caused the anthrax bacteria, which had been lying there frozen since a major outbreak in the 1940s, to wake up again and seek out new victims. Scientists had predicted that this scenario would occur when the permafrost began to thaw.

However, the greatest danger posed by the thawing of the permafrost lies in the enormous quantities of carbon that may be released into the atmosphere, triggering an irreversible warming. In this respect, we are on the threshold of nerve-racking times: a study led by Sarah Chadburn and published in *Nature Climate Change* says this can be avoided if we manage to restrict the temperature increase to 2.7 degrees Fahrenheit (1.5 degrees Celsius), but not if warming reaches 3.6 degrees Fahrenheit (2 degrees Celsius) above preindustrial levels. In that case, more than 40 percent of the area covered in permafrost could thaw, a total of 2.3 million square miles.[114]

But might there be ways of slowing this thaw?

— 23 —

CLIMATE HELP FROM THE ANIMAL KINGDOM

THE GOOD NEWS is that there *is*, in fact, something we can do. Neither the warming nor the changes in vegetation are inevitable, things that just happen regardless. We can hardly get these processes to stop entirely or reverse—at least not over the long term—but it is possible to slow the development and even reverse it at the local level. It isn't easy and it will require knowledge of how the effects will play out in the relevant locations. But still, it is worth trying and some people have already attempted it—successfully. And we won't have to do all of this alone, either. In part, we can leave it up to animals, to grazing animals. They can become our most important helpers in the battle to avoid climate catastrophe.

Reisadalen, a valley in northern Norway, leads up toward Finnmarksvidda and has been used for centuries as pastureland and a migration route for reindeer. Fifty years ago, long reindeer fences were built here to keep the reindeer herds in place. This has given scientists an opportunity to carry out systematic studies of how reindeer grazing affects the vegetation. Scientists from Sweden's Umeå University in particular have taken thorough measurements of a number of aspects, such as the density and nature of the vegetation, the humidity of the soil, temperature, and albedo. Since the readings were taken in the summer half of the year, we are dealing with summer albedo in this case—in other words, how much sunlight different types of vegetation reflect. The readings were carried out at an elevation of 1,600 to 2,300 feet above sea level. Up here in the north, that means you are roughly 330 feet above the tree line, with a mean annual temperature of 31 degrees Fahrenheit. The vegetation here is typically Arctic, consisting of brush, heather, and lichen, and the scientists classify it as tundra.

While the landscape on one side of the reindeer fence is almost untouched by grazing, the area on the other side is grazed by domesticated reindeer herds. So the scientists have a ready-made experiment, which stretches over a couple of miles and different types of tundra landscape. This setup makes it easy to measure the effect reindeer grazing has on different types of vegetation, compared with a nongrazed control landscape. The results show that the consequences of grazing point in a different direction than many might have expected. The scientists have investigated the climate effect in particular, because tundra is a landscape type that not only dominates large swaths of the planet but is also an important wild card when it comes to climate issues, owing to the enormous stores of carbon that lie frozen beneath it.

Without having done any proper research, many people have assumed that grazing animals are harmful to the environment, but in a climate context, it turns out that reindeer are among our best friends: they slow the warming of the tundra. Dutch scientist Mariska te Beest, leader of the Umeå University study, concluded: "In this study we therefore show that reindeer have a potential cooling effect on the climate. ... The estimated differences might appear small, but are large enough to have consequences for the regional energy balance." [115]

How can reindeer affect the climate? By eradicating the type of vegetation that reduces the albedo effect in both winter and summer, such as brush, and keeping the landscape open to vegetation such as reindeer lichen and grass, which is both their favorite food and the vegetation with the greatest climate-mitigating effect. The readings showed that the difference in "net radiation," which is determined by how much solar radiation is reflected, was so great between heavily grazed and lightly grazed areas that it was "equal to or higher, per unit of area, than the global atmospheric heating of 4.4 Wm^{-2} [watts per square meter] associated with a doubling of atmospheric CO_2." [116]

In other words, the reindeer grazing offset a doubling of the carbon dioxide content in the atmosphere! Admittedly, that is measured per unit of area, but considering what a vast area the tundra covers and how large the potential grazing area therefore is, the overall effect could be reasonably large, as long as reindeer grazing wasn't overdone (which could take a toll on the pale reindeer lichen, the vegetation offering the very best albedo effect).

But of course, the reindeer can't be expected to do all of the climate work by themselves. And, fortunately, they don't have to. The same job that reindeer do in their own landscape areas can be done by other grazing animals in landscapes with different types of vegetation. This has been demonstrated empirically over twenty years in a remarkable project in northeast Siberia. It may sound like some naive Disney story, but it may actually turn out that what will save us from climate catastrophe is animals, and lots of them at that.

THE ICE AGE PARK

THERE IS TWICE as much carbon and greenhouse gas in the permafrost as there currently is in the whole atmosphere. When the temperature rises, it is almost a law of nature that more and more of this carbon will flow out into the atmosphere in the form of carbon dioxide or, worse still, methane gas. And it may seem as if there is little we can do about it.

But somebody *is*, in fact, already trying to do something about it. And it seems to be working. Far northeast in Siberia, in the most desolate and inhospitable region imaginable, a persevering Russian has dedicated more than twenty years to a project that might have seemed like a fantasy at the outset but that has increasingly been attracting attention from serious scientific quarters in recent years, including *Science*, one of the world's two most important scientific magazines.[117]

This man, Sergey Zimov, wants to stop the melting in the tundra, and with it the emission of catastrophic amounts of greenhouse gases, by restoring the tundra to what it once used to be: a steppe landscape where large herds of animals wandered and grazed. The change that this brings with it actually has major climate effects: it causes the melting of the permafrost to reverse, while notably strengthening the albedo effect. This does not only happen because the snow cover becomes more constant but also because, even without snow cover, grass reflects sunlight better than shrubs, moss, and forest. That may make it possible to slow the rise in temperature even on a global scale—if the project can be expanded to cover larger parts of the tundra. If this works—and the experiment so far shows that it does—this may be the most important of all the climate actions we can undertake.

Zimov's project has implications far beyond what he can achieve in Siberia. It will show that we can deal with the climate threat by beginning to think a bit beyond emissions alone. As we can learn from history, the main influences on the climate are the feedback processes for which nature itself is responsible (admittedly hastened by humans in some cases). What happens in the interface between land and atmosphere, sea and sky, in a complex interplay of chemical, biochemical, and physical processes, determines whether the carbon balance will tip one way or the other. The strength of the albedo effect is also an important factor—how much solar energy is reflected, and therefore doesn't contribute to warming. Zimov's project has great relevance for all of these paramount concerns.

It was in 1988 that Sergey Zimov came to Chersky, a town by the Kolyma River in northeast Siberia so far off the beaten track that it is four to five hours by plane to the nearest largish city, Yakutsk. This is one of the coldest regions in the northern hemisphere, with winter temperatures as cold as −76 degrees Fahrenheit, or worse.

As a scientist, Zimov noticed a paradox: whereas life here in the north was now characterized by few animals and poor biological diversity, the thawed permafrost revealed that things had once been different. The fossils of numerous animals turned up here, including large, extinct beasts like mammoths. What had happened to all these animals?

A common explanation for the disappearance of mammoths and other megafauna has been climate change that occurred after the last ice age. It became too warm either for the animals or for the vegetation that sustained them. But Zimov came to the conclusion that this couldn't be right. The fossils showed that these animals had survived equally dramatic climate changes in previous periods. So why not this time? Why did they disappear around 13,000 years ago?

The arrival of human beings was probably the main reason. Equipped with effective hunting technology and hunting methods they had learned during the ice age, they had now become such efficient hunters that it was not long before they had wiped out the entire population of the largest animals (whose disadvantage was that they reproduced slowly). It was the same story around the entire globe: Wherever humans appeared—in Europe, Asia, Australia, America—the megafauna were rapidly eradicated, even though there were very different animal species in these different parts of the world. The only place the megafauna held their own was in Africa. One possible reason why the big animals survived there was that they had adapted to a life alongside human ancestors for millions of years and "knew" how to avoid these predators. But there may also have been other reasons to do with the specific landscape and species.

Zimov's project began as a test to show that it wasn't the climate but human hunting that had wiped out the megafauna. If he could get animal herds to survive here on the tundra at today's temperatures, they would also have been able to survive 13,000 years ago. But the project gradually developed into something more, a climate project, because Zimov saw that bringing animals back to the tundra was also doing something for the climate.

Those who live in areas where there are sheep know that grazing animals create and maintain their own landscape. Where grass is eaten, grass grows, and the grass landscape spreads along with the herbivores. This is what has shaped much of the Norwegian landscape, as in the country's most important sheep-farming district, Dalane in western Norway. In large areas here there is grass, grass, and nothing but grass in the places where the sheep spend most of the year grazing (and some sheep stay outside all year round). This landscape has become that way because by going about grazing there, the sheep halt any potential growth of shrubs, and fertilize the grass with their own droppings.

This was also how it used to be in Siberia until the humans came. Mammoths, Arctic antelopes (saiga), reindeer, wild horses, musk oxen, elk, and other herbivores grazed, while predators such as wolves, bears, and Siberian tigers "trimmed" the populations, kept them in shape, and ensured that they didn't spend too much time in one place (that way, areas rarely became overgrazed).

When human beings arrived and eventually wiped out the population of herbivores, the landscape began to change. Today, there are marshes, shrubs, and mossy forest where once there was grass. And this isn't the only change.

The snow cover has also altered. The great animal herds that previously existed behaved the same way as the Sami reindeer herds do in winter: they trampled the snow and dug through it to find food (reindeer lichen in the reindeer's case). And in so doing, these animals reduced the insulating effect of the snow. That enabled the part of the permafrost that had thawed in summer to freeze again. In this way, the animal herd helped maintain the permafrost deep in the ground, while only the upper layer thawed in the summer, allowing for rapid growth in the midnight sun.

But when the animals are gone (apart from the occasional reindeer herd), the snow is left to lie in peace and the earth doesn't freeze as far down as it otherwise would. As a result, the thawing happens faster, which is why, these days, we find that previously frozen mammoths—and bacteria—are reappearing. But whereas the mammoths can't be resurrected, despite being otherwise pretty much intact, the bacteria can come back to life after thousands of years in hibernation. As if this scenario wasn't worrying enough in itself, the thawing can also release climate gases in vast quantities, as discussed in the previous chapter.

So the project Zimov set in motion, later aided by his son, Nikita, and interested scientists from all over the globe, was to rebuild a fauna of herbivores on the landscape. He started with horses, a special breed that copes well with the cold. After a while they also introduced musk oxen, elk, and other animals, although this has all happened step by step, under controlled conditions. Zimov has had to set up fences to protect the animals from poachers.

Ideally, Zimov would also like to have mammoths in his "ice age park," Pleistocene Park. And efforts are now being made, in fact, to recreate mammoths, because some roughly intact mammoth DNA has been found frozen in the tundra; based on that, scientists hope to produce new animals with the help of their relative the elephant.[118] Zimov also wants to introduce predators such as wolves and Siberian tigers.

To start off with, in 1988, it was still possible for Zimov to get public funding for his project. But when the Soviet Union collapsed, there was no longer any money to be had. Zimov had to seek support from abroad. The project has gradually attracted major international attention, but it still lacks the funding it needs to realize all its plans.

Even so, the results so far are impressive. Zimov and his assistants have managed to introduce animals—herbivores—and even though this initially caused problems, they have managed to get them to live there. And so Zimov has proved that it wasn't the climate that finished off the populations 13,000 years ago, because the climate was roughly the same then as now. The climate effect thus far has also turned out to be as Zimov forecast: the herbivores have a cooling effect on the soil, thereby helping slow the melting of the permafrost. Zimov's son, Nikita, who now leads the research station, says that the animals make a dramatic difference to the snow's insulating effect.

"We have placed the temperature-measuring probes in the soil, where the animals stampeded the snow, and on landscapes void of animals. Observations and data collected at the park reveal that even when outside air temperature is -40 C [-40 degrees Fahrenheit], the temperature of the ground under the snow that is void of animals is only -5 C [23 degrees Fahrenheit]. But in areas where animals stampeded the snow, the temperature dropped down to -30 C [-22 degrees Fahrenheit]."[119]

The point is, therefore, that untouched snow, without animals, insulates the earth so well that winter cold cannot penetrate it. Consequently, the "active" permafrost layer, the upper part which melts in summer, does not refreeze as much in the winter. The animals' presence, their trampling and digging in the snow, reduces the insulating effect, ensuring that the tundra refreezes in winter—thereby preserving the permafrost. And the enormous amounts of carbon concealed by the permafrost—the climate bombs on the tundra—are not released. Or not as quickly as they would have been, at least.

LAST DANCE?
(IS THERE A FUTURE FOR
THE CRYOSPHERE?)

*The Inuit hunter who falls through the depleting
and unpredictable sea ice is connected to
the cars we drive, the industries we rely upon, and
the disposable world we have become.*

SHEILA WATT-CLOUTIER, former president of the Inuit
Circumpolar Council, testimony to the US Senate, 2005

S EEN FROM THE outside by an observer with plenty of time
on their hands, the Earth has been performing a regular
dance for many millions of years: the white caps on the
poles in north and south have shrunk and then expanded again
in a regular pattern known to us earth-dwellers as ice ages.

This pattern has lasted for as long as there have been humans here, and far back in the time of our forebears. And the poles aren't the only places that have had white patches. Scattered around the globe, wherever there have been high mountains, the white Kingdom of Frost has also had its outposts.

But now something doesn't make sense. The dance should actually be on its way back to the white, to a new ice age. Instead, the white caps are vanishing faster than ever before. Seen from here down below, it seems as if a melting is underway. The sea ice is diminishing and so are the mountain glaciers, along with the snow cover that has spread over large expanses of the land surface every winter. Even the ice sheets of Greenland and Antarctica have begun to crack up.

We do not know whether this will continue. The geophysical mechanisms that control this are only partially clear. We do not know when—or whether—the Earth's systems will pass the tipping point that will set us on an unstoppable course for a process of warming. But what we do know is that without the cryosphere, life on Earth will become difficult for us humans. Hundreds of millions will be left without water, heat waves and forest fires will make vast areas uninhabitable, the weather will become steadily more extreme, and the ocean will rise to levels well above the places where most large cities and population centers stand today.

But it is not yet certain that this will happen. That is why even the small actions of people like Zimov, with his Pleistocene Park, and the cryoactivists, with their battle for the glaciers, may be just what is needed to prevent it. Then the Kingdom of Frost can strike back, just as it has so many times in the past.

NOTES

---

[1] "Cryo" comes from the Greek *kryo*, "cold, frost"—so the cryosphere is the part of the world where water is frozen (in the form of ice, snow, or permafrost).

[2] "Norge i rødt, hvitt og blått" [Norway in red, white, and blue], lyrics written in 1941 by Finn Bø, Bias Bernhoft, and Arild Feldborg.

[3] Thomas Alsgaard to Norwegian newspaper *VG*, July 11, 2017.

[4] John Maynard Smith and Eörs Szathmáry, *The Origins of Life: From the Birth of Life to the Origin of Language* (Oxford: Oxford University Press, 1999).

[5] Nick Lane, *The Vital Question: Why Is Life the Way It Is?* (London: Profile Books, 2015).

[6] Examples of panspermia in popular literature: Jack Finney, *The Body Snatchers* (1955); Fred Hoyle, *The Black Cloud* (1957); Michael Crichton, *The Andromeda Strain* (1969). In science: Francis Crick and Leslie Orgel, "Directed Panspermia," *Icarus* 19, no. 3 (1973): 341–48.

[7] Rachel Courtland, "'Water Bears' Are First Animals to Survive Space Vacuum," *New Scientist*, September 8, 2008, www.newscientist.com/article/dn14690-water-bears-are-first-animal-to-survive-space-vacuum/.

[8] Kathrin Altwegg, Hans Balsiger, Akiva Bar-Nun, Jean-Jacques Berthelier, Andre Bieler, Peter Bochsler, Christelle Briois, et al., "Prebiotic

Chemicals—Amino Acid and Phosphorus—in the Coma of Comet 67P/ Churyumov-Gerasimenko," *Science Advances* 2, no. 5 (May 27, 2016): e1600285, doi:10.1126/sciadv.1600285.

9 Facts about albedo largely drawn from Shawn J. Marshall, *The Cryosphere* (Princeton, NJ: Princeton University Press, 2012).

10 Nordahl Grieg's poem "Morgen over Finnmarksvidden" [Morning on Finnmarksvidda] is from the collection *Norge i våre hjerter* [Norway in our hearts], 1929.

11 Sophus Tromholt, *Under the Rays of the Aurora Borealis: In the Land of the Lapps and Kvæns*, Vol. 1 (London: Sampson Low, Marston, Searle, and Rivington, 1885), 94–95, https://archive.org/details/underraysaurora01 siewgoog/page/n8.

12 Tor Åge Bringsværd, *Vår gamle gudelære. 1. En kjempe så stor som hele verden* [Our old mythology: A giant as big as the whole world] (Oslo: Gyldendal Norsk Forlag, 1985), 7–8.

13 Bringsværd, *Vår gamle gudelære*, 10–11.

14 Bringsværd, *Vår gamle gudelære*, 16–19.

15 Fridtjof Nansen, *Farthest North: Being the Record of a Voyage of Exploration of the Ship "Fram" 1893–96*, Vol. 1 (1897, Project Gutenberg, EBOOK #30197, 2009), 280, http://www.gutenberg.org/files/30197/30197-h/30197-h. htm.

16 Grieg, "Morgen over Finnmarksvidden."

17 Nils Jernsletten, "Sami Traditional Terminology: Professional Terms concerning Salmon, Reindeer and Snow," in *Sami Culture in a New Era: The Norwegian Sami Experience*, ed. Harald Gaski (Karasjok, Norway: Davvi Girji, 1997), 86–108.

18 Ole Henrik Magga, "Diversity in Saami Terminology for Reindeer, Snow, and Ice," *International Social Science Journal* 58, no. 187 (2006): 25–34, doi:10.1111/j.1468-2451.2006.00594.x.

19 Inger Marie Gaup Eira, Christian Jaedicke, Ole Henrik Magga, Nancy G. Maynard, and Svein D. Mathiesen, "Traditional Sámi Snow Terminology and Physical Snow Classification—Two Ways of Knowing," *Cold Regions Science and Technology*, 85 (January 2013): 117–30, doi:10.1016/ j.coldregions.2012.09.004.

20 Yngve Ryd, *Snö: Renskötaren Johan Rassa berättar* [Snow: An account by reindeer herder Johan Rassa] (Stockholm: Natur & Kultur, 2007), 7.

[21] Ryd, *Snö*, 8.

[22] In this discussion, the Lulesamisk terms themselves follow Ryd, *Snö*. The accompanying information is drawn not only from Ryd but also from the sources listed in notes 17–19, as well as from my own personal knowledge.

[23] Ryd, *Snö*, 36.

[24] Ryd, *Snö*, 42.

[25] About Esmark: Jamie Woodward, *The Ice Age: A Very Short Introduction* (Oxford: Oxford University Press, 2014).

[26] About Agassiz: Woodward, *The Ice Age*.

[27] Louis Agassiz, quoted by Doug Macdougall, *Frozen Earth: The Once and Future Story of Ice Ages* (Berkeley: University of California Press, 2013), 36.

[28] "Snowball Earth": The first discoveries were made by J. Thomson in Scotland in 1871 and Hans Reusch in northern Norway in 1891. Joe Kirschvink's article introduced the term: Joseph L. Kirschvink, "Late Proterozoic Low-Latitude Global Glaciation: The Snowball Earth," in *The Proterozoic Biosphere: A Multidisciplinary Study,* eds. J. William Schopf and Cornelius Klein (Cambridge: Cambridge University Press, 1992), 51–52.

[29] Tim Lenton and Andrew Watson, *Revolutions That Made the Earth* (Oxford: Oxford University Press, 2011).

[30] Jim Roche, quoted by Jeremy Miller, "The Dying Glaciers of California," *Earth Island Journal*, June 1, 2013, http://www.earthisland.org/journal/index.php/magazine/entry/the_dying_glaciers_of_california/.

[31] Greg Stock and Robert Anderson, "Yosemite's Melting Glaciers," Final Report (San Francisco: Yosemite Conservancy, December 2012).

[32] Jeffery P. Schaffer, *Yosemite National Park: A Natural-History Guide to Yosemite and Its Trails*, 4th ed. (Berkeley, CA: Wilderness Press, 1999).

[33] Eric Holthaus, "The Thirsty West: 10 Percent of California's Water Goes to Almond Farming," *Slate*, May 14, 2014, http://www.slate.com/articles/technology/future_tense/2014/05/_10_percent_of_california_s_water_goes_to_almond_farming.html.

[34] Jacqueline Ho, Ingrid Maradiaga, Jamika Martin, Huyen Nguyen, and Linh Trinh, "Almond Milk vs. Cow Milk: Life Cycle Assessment," Environment 159 class paper, Institute of the Environment and Sustainability, UCLA, June 2, 2016, https://www.ioes.ucla.edu/wp-content/uploads/cow-vs-almond-milk-1.pdf (no longer available).

[35] According to the CIA World Factbook 2017 (https://www.cia.gov/library/publications/the-world-factbook/), the two biggest wheat exporters were the US (14.8 percent of the world total) and Canada (12.4 percent). The US also accounted for 35.9 percent of corn exports.

[36] L. A. Vincent, X. Zhang, R. D. Brown, Y. Feng, E. Mekis, E. J. Milewska, H. Wan, et al., "Observed Trends in Canada's Climate and Influence of Low Frequency Variability Modes," *Journal of Climate* 28 (June 1, 2015): 4545–60, doi:10.1175/JCLI-D-14-00697.1.

[37] IPCC, *Climate Change 2013: The Physical Science Basis. Contribution of Working Group I to the Fifth Assessment Report of the Intergovernmental Panel on Climate Change,* eds. T. F. Stocker, D. Qin, G.-K. Plattner, M. Tignor, S. K. Allen, J. Boschung, A. Nauels, et al. (Cambridge and New York: Cambridge University Press, 2013), https://www.ipcc.ch/report/ar5/wg1/#.

[38] Barrie Bonsal, Environment and Climate Change Canada, personal communication, December 14, 2018; and John Pomeroy, quoted by Erin Collins, "It's Not Impossible: Western Canada's Risk of Water Shortages Rising," *CBC News,* March 14, 2018, https://www.cbc.ca/news/canada/calgary/africa-capetown-water-shortage-drought-canada-rockies-glacier-1.4564616.

[39] Judith D. Schwartz, *Water in Plain Sight: Hope for a Thirsty World* (New York: St. Martin's Press, 2016), 42.

[40] Christopher White, *The Melting World: A Journey across America's Vanishing Glaciers* (New York: St. Martin's Press, 2013), 103–4.

[41] Bob Berwyn, "Unabated Global Warming Threatens West's Snowpack, Water Supply," *InsideClimate News,* June 7, 2016, https://insideclimatenews.org/news/07062016/unabated-global-warming-threatens-west-snowpack-water-rocky-mountains-sierra-nevada-drought.

[42] Cited in White, *The Melting World,* 37: "I'm not sure the glaciers have twenty-two years left"; White followed Fagre in his work in Glacier National Park for five years.

[43] Glaciers in Canada: Robert William Sandford, *Our Vanishing Glaciers: The Snows of Yesteryear and the Future Climate of the Mountain West* (Victoria, BC: RMB, 2017).

[44] White, *The Melting World.*

[45] Hina Alam, "80% of Mountain Glaciers in Alberta, B.C. and Yukon Will Disappear within 50 Years: Report," *CBC News,* December 27, 2018,

https://www.cbc.ca/news/canada/british-columbia/western-glaciers-disappear-50-years-1.4959663.

[46] Jeb Bell, Glenn Tootle, Larry Pochop, Greg Kerr, and Ramesh Sivanpillai, "Glacier Impacts on Summer Streamflow in the Wind River Range, Wyoming," *Journal of Hydrologic Engineering* 17, no. 4 (April 2012): 521–27, doi:10.1061/(ASCE)HE.1943-5584.0000469.

[47] Hanspeter Liniger, Rolf Weingartner, and Martin Grosjean, *Mountains of the World: Water Towers for the 21st Century*, prepared for the United Nations Commission on Sustainable Development (Berne: Institute of Geography, University of Berne, and Swiss Agency for Development and Cooperation, 1998), http://lib.icimod.org/record/10237/files/146.pdf.

[48] Sandford, *Our Vanishing Glaciers*, 205.

[49] Andreas Staeger, "Handmade for Your Feet," *TOP Magazin*, 2018, 20–26, http://top-magazin.ch/2018/.

[50] Andreas Bauder, ed., *The Swiss Glaciers 2013/14 and 2014/15*, Glaciological Report No. 135/136 (Cryospheric Commission [EKK], 2017), doi:10.18752/glrep135-136.

[51] Bauder, "The Swiss Glaciers."

[52] Jeannette Nötzli, Rachel Lüthi, and Benno Staub, eds., *Permafrost in Switzerland 2010/2011 to 2013/2014*, Glaciological Report (Permafrost) No. 12-15, Swiss Permafrost Monitoring Network (Cryospheric Commission of the Swiss Academy of Sciences, 2016), https://naturalsciences.ch/service/publications/82035-permafrost-in-switzerland-2010-2011-to-2013-2014.

[53] *I Elvegudinnens Rike* [In the realm of the river goddess], directed and written by Bjørn Vassnes (Oslo/Bergen: NRK/Univisjon, 1994), television series.

[54] Yogesh Kumar and Tapan Susheel, "On Uma Bharti's Direction, Experts to Probe if Kailash is Ganga's Source as Hindu Mythology Says," *Times of India*, October 30, 2015, https://timesofindia.indiatimes.com/india/On-Uma-Bhartis-direction-experts-to-probe-if-Kailash-is-Gangas-source-as-Hindu-mythology-says/articleshow/49575226.cms.

[55] Nansen, *Farthest North*, 2.

[56] A. H. Winsnes, ed., *Nansen's røst, artikler og taler av Fridtjof Nansen I* [Nansen's voice: Articles and speeches by Fridtjof Nansen] (Norway: Jacob Dybwads Forlag, 1942), 181.

[57] Nansen, *Farthest North*, 280–81.

[58] Hjalmar Johansen, *Med Nansen mot Nordpolen* [With Nansen to the North Pole] (1898; Oslo: Kagge, 2007), 184. Page reference is to the 2007 edition.

[59] Richard Boyle, cited in Lenton and Watson, *Revolutions That Made the Earth.*

[60] Lenton and Watson, *Revolutions That Made the Earth*, 282.

[61] *March of the Penguins* (original title *La marche de l'empereur*), directed by Luc Jacquet (France, 2005), film.

[62] Kjetil Lysne Voje, "Tempo Does Not Correlate with Mode in the Fossil Record," *Evolution* 70, no. 12 (December 2016): 2678–89, doi:10.1111/evo.13090.

[63] Tardigrades: see note 7.

[64] The Cambrian explosion: The best-known description of this is in Stephen Jay Gould, *Wonderful Life: The Burgess Shale and the Nature of History* (New York: W. W. Norton, 1989).

[65] Elisabeth S. Vrba, "The Pulse That Produced Us," *Natural History* magazine 102, no. 5 (May 1993): 47–51.

[66] Mark A. Maslin, Chris M. Brierley, Alice M. Milner, Susanne Shultz, Martin H. Trauth, and Katy E. Wilson, "East African Climate Pulses and Early Human Evolution," *Quaternary Science Reviews* 101 (October 2014): 1–17, doi:10.1016/j.quascirev.2014.06.012.

[67] Richard Potts, "Evolution and Climate Variability," *Science* 273, no. 5277 (August 16, 1996): 922–23, doi:10.1126/science.273.5277.922.

[68] Steven Mithen, *After the Ice: A Global Human History 20,000–5000 BC* (London: Phoenix, 2003).

[69] Abu Hureyra: described in Brian Fagan, *The Long Summer: How Climate Changed Civilization* (New York: Basic Books, 2004). How climatic changes have influenced human history was perhaps first written about by Brian Fagan, professor emeritus at the University of California, Santa Barbara, in books including *The Long Summer* as well as *Floods, Famines, and Emperors: El Niño and the Fate of Civilizations* (New York: Basic Books, 1999) and *The Little Ice Age: How Climate Made History, 1300–1850* (New York: Basic Books, 2000).

Moore's own account in A. M. T. Moore, G. C. Hillman, and A. J. Legge, *Village on the Euphrates: From Foraging to Farming at Abu Hureyra* (Oxford: Oxford University Press, 2000).

[70] Manfred Heun, interview in forskning.no (January 18, 2012). About the domestication of einkorn: Manfred Heun, Ralf Schäfer-Pregl, Dieter Klawan, Renato Castagna, Monica Accerbi, Basilio Borghi, and Francesco Salamini, "Site of Einkorn Wheat Domestication Identified by DNA Fingerprinting," *Science* 278, no. 5341 (November 14, 1997): 1312–14, doi:10.1126/science.278.5341.1312. About beer and religious cults in Göbekli Tepe: Oliver Dietrich, Manfred Heun, Jens Notroff, Klaus Schmidt, and Martin Zarnkow, "The Role of Cult and Feasting in the Emergence of Neolithic Communities: New Evidence from Göbekli Tepe, South-Eastern Turkey," *Antiquity* 86, no. 333 (September 2012): 674–95, doi:10.1017/S0003598X00047840.

[71] On the transition from shamanism to organized religion see, among others, Bjørn Vassnes, *Sjelens sult* [The soul's hunger] (Tromsø, Norway: Margbok, 2009) and Ara Norenzayan, *Big Gods: How Religion Transformed Cooperation and Conflict* (Princeton, NJ: Princeton University Press, 2013).

[72] Marshall Sahlins, "Notes on the Original Affluent Society," in *Man the Hunter*, eds. Richard B. Lee and Irven DeVore (Chicago: Aldine Publishing Company, 1968), 85–89.

[73] Gregory Clark, *A Farewell to Alms—A Brief Economic History of the World* (Princeton, NJ: Princeton University Press, 2007); Ian Morris, *The Measure of Civilization: How Social Development Decides the Fate of Nations* (Princeton, NJ: Princeton University Press, 2013).

[74] Clark, *A Farewell to Alms*, 1.

[75] Clark, *A Farewell to Alms*, 1.

[76] Matthias Foss, *Justedalens kortelige Beskrivelse* [A short description of Justedalen] (1750; Jostedal historielag, 2009), 10, http://www.historielaget.jostedal.no/?page_id=190. Page reference is to the 2009 edition.

[77] Foss, *Justedalens kortelige Beskrivelse*, 10.

[78] Foss, *Justedalens kortelige Beskrivelse*, 10.

[79] On the 1657/58 war in Denmark: Thomas Roth, "Den snöige nord" [The snowy north], appendix to Erik Durschmied, *Vädrets makt* [The weather factor] (Stockholm: Prisma, 2002).

[80] James Daschuk, *Clearing the Plains: Disease, Politics of Starvation, and the Loss of Aboriginal Life* (Regina, SK: University of Regina Press, 2013).

[81] The Norse colony on Greenland: Jared Diamond, *Collapse: How Societies Choose to Fail or Succeed* (New York: Penguin, 2005).

82 Daschuk, *Clearing the Plains*, 5.

83 In *The Human Planet: How We Created the Anthropocene* (London: Pelican, 2018), Simon L. Lewis and Mark A. Maslin present a new argument: that the coldest period of the Little Ice Age may have been triggered by the European invasion of the Americas. This led to great losses in the population, which left large agricultural areas unworked and led to a huge regrowth of forests, resulting in a substantial uptake of CO_2, and cooling.

84 Peter Høeg, *Smilla's Sense of Snow*, trans. Tiina Nunnally (New York: Farrar, Straus and Giroux, 1993), 402.

85 Arctic Monitoring and Assessment Programme, *Snow, Water, Ice and Permafrost in the Arctic: Summary for Policy-Makers*, 2017, https://www.amap.no/documents/doc/Snow-Water-Ice-and-Permafrost.-Summary-for-Policy-makers/1532.

86 Harald Steen to Norwegian broadcaster NRK, October 15, 2017.

87 On the Jakobshavn Glacier, see, for example: Jane J. Lee, "Greenland Glacier Races to Ocean at Record Speed," *National Geographic*, February 14, 2014, https://news.nationalgeographic.com/news/2014/02/140204-greenland-glacier-iceberg-speed-climate-change-science/; and Chris Mooney, "Scientists Just Uncovered Some Troubling News about Greenland's Most Enormous Glacier," *Washington Post*, April 11, 2017, https://www.washingtonpost.com/news/energy-environment/wp/2017/04/11/scientists-just-uncovered-some-troubling-news-about-greenlands-most-enormous-glacier/.

88 *Om Golfstrømmen skulle stanse* [If the Gulf Stream stopped], directed and written by Bjørn Vassnes (Oslo/Bergen: NRK/Univisjon, 2001), television program.

89 Douglas Fox, "The Larsen C Ice Shelf Collapse Is Just the Beginning—Antarctica Is Melting," *National Geographic*, June 14, 2017, https://www.nationalgeographic.com/magazine/2017/07/antarctica-sea-level-rise-climate-change/.

90 Leah Rosenbaum, "The Giant Iceberg That Broke from Antarctica's Larsen C Ice Shelf Is Stuck," *Science News*, July 23, 2018, https://www.sciencenews.org/article/giant-iceberg-broke-antarctica-larsen-c-ice-shelf-stuck.

91 NASA Earth Observatory, "New Studies Get to the Bottom of Antarctic Melting," January 17, [2017], https://earthobservatory.nasa.gov/images/89016/new-studies-get-to-the-bottom-of-antarctic-melting.

[92] Stephen Rich Rintoul, Alessandro Silvano, Beatriz Pena-Molino, Esmee van Wijk, Mark Rosenberg, Jamin Stevens Greenbaum, and Donald D. Blankenship, "Ocean Heat Drives Rapid Basal Melt of the Totten Ice Shelf," *Science Advances* 2, no. 12 (December 16, 2016): e1601610, doi:10.1126/sciadv.1601610.

[93] Lide Tian, Tandong Yao, Yang Gao, Lonnie Thompson, Ellen Mosley-Thompson, Sher Muhammad, Jibiao Zong, et al., "Two Glaciers Collapse in Western Tibet," *Journal of Glaciology* 63, no. 237 (February 2017): 194–97, doi:10.1017/jog.2016.122.

[94] Tandong Yao, Lonnie Thompson, Wei Yang, Wusheng Yu, Yang Gao, Xuejun Gao, Xiaosin Yang, et al., "Different Glacier Status with Atmospheric Circulations in Tibetan Plateau and Surroundings," *Nature Climate Change* 2 (July 15, 2012): 663–67, doi:10.1038/nclimate1580.

[95] Thomas Jacob, John Wahr, W. Tad Pfeffer, and Sean Swenson, "Recent Contributions of Glaciers and Ice Caps to Sea Level Rise," *Nature* 482 (February 23, 2012): 514–18, doi:10.1038/nature10847.

[96] Tandong Yao, quoted by Jane Qiu, "Tibetan Glaciers Shrinking Rapidly," *Nature* News, July 15, 2012, doi:10.1038/nature.2012.11010.

[97] IRIN website, "Himalayan Glaciers Melting More Rapidly," July 20, 2012, http://www.irinnews.org/report/95917/climatechange-himalayan-glaciers-melting-more-rapidly.

[98] IRIN website, "Himalayan Glaciers Melting More Rapidly."

[99] GLACINDIA study: http://uni.no/en/uni-climate/climate-variability/glacindia-water-related-effects-of-changes-in-glacier-mass-balance-and-river-runoff-inwestern-hima/.

[100] Markus Engelhardt, Al. Ramanathan, Trude Eidhammer, Pankaj Kumar, Oskar Landgren, Arindan Mandal, and Roy Rasmussen, "Modelling 60 Years of Glacier Mass Balance and Runoff for Chhota Shigri Glacier, Western Himalaya, Northern India," *Journal of Glaciology* 63, no. 240 (August 2017): 618–28, doi:10.1017/jog.2017.29.

[101] Jane Qiu, "Investigating Climate Change the Hard Way at Earth's Icy 'Third Pole,'" *Scientific American*, June 2, 2016, https://www.scientificamerican.com/article/investigating-climate-change-the-hard-way-at-earth-s-icy-third-pole/.

[102] International Centre for Integrated Mountain Development (ICIMOD), "Glacial Lakes and Glacial Lake Outbursts in Nepal," March 2011, http://www.icimod.org/dvds/201104_GLOF/reports/final_report.pdf.

[103] Scientific India website, "Glacial Lakes Threaten Indian Himalayan Dams," August 31, 2016, http://www.scind.org/341/Mindblower/glacial-lakes-threaten-indian-himalayan-dams.html.

[104] Jorge Daniel Taillant, *Glaciers: The Politics of Ice* (Oxford: Oxford University Press, 2015).

[105] On developments in the Andes: Jorge Daniel Taillant, "Government Leaks on Glacier Impacts Complicate Several Mining Projects in Argentina," Center for Human Rights and Environment (CHRE) website, December 27, 2016, http://center-hre.org/?p=15904; Jorge Daniel Taillant, "Argentina Federal Government Recognizes Glaciers in Mining Areas," CHRE website, January 26, 2017, http://center-hre.org/?p=15998; Antonio De la Jara, "Chile Environmental Court Orders Barrick to Close Pascua-Lama Gold Mine," October 12, 2018, Reuters, https://www.reuters.com/article/us-barrick-gold-chile/chile-environmental-court-orders-barrick-to-close-pascua-lama-gold-mine-idUSKCN1MM2LA.

[106] Taillant, *Glaciers*. Norphel's "artificial glaciers" in Ladakh (which is the Indian part of the Tibetan plateau and therefore very dry) have been developed further by Sonam Wangchuk, also an engineer. He has found a way to make ice stupas, large towers of ice formed like Tibetan stupas. How these are made can be seen on a YouTube video: https://www.youtube.com/watch?v=kptgonELjoo.

[107] Dana Desonie, *Polar Regions: Human Impacts* (New York: Chelsea House, 2008).

[108] *Siberian Times*, "Now the Proof: Permafrost 'Bubbles' Are Leaking Methane 200 Times above the Norm," July 22, 2016, https://siberiantimes.com/ecology/casestudy/news/n0681-now-the-proof-permafrost-bubbles-are-leaking-methane-200-times-above-the-norm/.

[109] *Siberian Times*, "7,000 Underground Gas Bubbles Poised to 'Explode' in Arctic," March 20, 2017, https://siberiantimes.com/science/casestudy/news/n0905-7000-underground-gas-bubbles-poised-to-explode-in-arctic/.

[110] *Siberian Times*, "Big Bang Formed Crater Causing 'Glow in Sky': Explosion Was Heard 100 km Away," June 7, 2016, https://siberiantimes.com/science/casestudy/news/n0676-big-bang-formed-crater-causing-glow-in-sky-explosion-was-heard-100-km-away/.

[111] Charles J. Hanley, "Climate Trouble May Be Bubbling Up in Far North," Phys.org, August 30, 2009, https://phys.org/news/2009-08-climate-north.html.

[112] Hanley, "Climate Trouble."

[113] Michaeleen Doucleff, "Anthrax Outbreak in Russia Thought to Be Result of Thawing Permafrost," NPR, August 3, 2016, https://www.npr.org/sections/goatsandsoda/2016/08/03/488400947/anthrax-outbreak-in-russia-thought-to-be-result-of-thawing-permafrost.

[114] S. E. Chadburn, E. J. Burke, P. M. Cox, P. Friedlingstein, G. Hugelius, and S. Westermann, "An Observation-Based Constraint on Permafrost as a Function of Global Warming," *Nature Climate Change* 7 (April 10, 2017): 340–44, doi:10.1038/nclimate3262.

[115] Mariska te Beest, Judith Sitters, and Johan Olofsson, "Reindeer Grazing Increases Summer Albedo by Reducing Shrub Abundance in Arctic Tundra," *Environmental Research Letters* 11, no. 12 (December 22, 2016): 125013, doi:10.1088/1748-9326/aa5128.

[116] Te Beest et al., "Reindeer Grazing Increases Summer Albedo."

[117] Sergey A. Zimov, "Pleistocene Park: Return of the Mammoth's Ecosystem," *Science* 308, no. 5723 (May 6, 2005): 796–98, doi:10.1126/science.1113442.

[118] Britt Wray, *Rise of the Necrofauna: The Science, Ethics, and Risks of De-Extinction* (Vancouver, BC: Greystone Books, 2017).

[119] Nikita Zimov, quoted by Damira Davletyarova, "The Zimovs: Restoration of the Mammoth-Era Ecosystem, and Reversing Global Warming," *Ottawa Life Magazine*, February 11, 2013, http://www.ottawalife.com/article/the-zimovs-restoration-of-the-mammoth-era-ecosystem-and-reversing-global-warming.

INDEX

Abu Hureyra, 150–52

Africa, 134–35, 139, 141, 142, 144–46, 189, 209–10

Agassiz, Louis, 53–55

agriculture, 150–58; cash crops, 164–65; civilization from, 155, 156–57; climate change and development of, 153–55; Little Ice Age impacts, 162–63; quality of life under, 157–58; transition to, 148–49, 150–53

albedo effect: climate effects, 21–22, 59; explanation of, 23–24; grazing animals and, 205, 206, 208; reindeer lichen and, 207; snowball Earth and, 60; snow cover and sea ice, 129–30, 175–76

Aletsch Glacier, 90, 94

almonds, 76

Alps, 88–96; 2018 snow issues, 88–89, 90; glacier changes, 91, 93–95; meltwater changes, 96; mountain dairy farming, 91; permafrost changes and landslides, 92–93, 95–96; skiing in, 89–90; snow cover changes, 91–92, 93; tourism in, 90

Alsgaard, Thomas, 3

Altwegg, Kathrin, 13

Amundsen, Roald, 115, 118–20, 123–25, 126, 137, 170

Andersen, Hans Christian, 30, 31

Andes, 6, 69, 189, 193–95

animals. *See* grazing animals; life

Antarctica, 118–27; animal life on, 122; climate change and, 177–80; as coldest continent, 121–22; exploration of, 41, 120, 122–23; ice on, 64, 120, 122–23; ice shelves, 177–78; overview, 120–21; race for South Pole, 118–20, 123–25; sea ice, 129;